Introduction

Today it may come as something of a surprise to see so many countries opting to purchase the F-35 stealth fighter from Lockheed Martin. It has, after all, been more than three decades since the last such combat aircraft sales bonanza and times have changed somewhat.

But America has famously been the arsenal of democracy since the Second World War – supplying the latest, or nearly the latest, weapons technology to friendly nations across the globe. And its most high profile defence export has always been fighter aircraft.

Some have seen relatively modest use beyond their native land; few foreign users adopted the F-80 Shooting Star, the F2H-3 Banshee or the F9F in either Panther or Cougar form. Even newer types such as the F-100 Super Sabre, F-101 Voodoo, F-102 Delta Dagger, F-8 Crusader and F-14 Tomcat were rarely operated by America's allies.

Yet seemingly for every 'miss' there was a 'hit' so spectacularly successful in terms of production numbers and individual air force users that it put jet fighters produced by every other nation outside the Soviet Union in the shade.

The first such success was the Republic F-84G Thunderjet, which equipped countries across Europe and beyond during the early 1950s. This was swiftly followed by the F-84F Thunderstreak but both were soon eclipsed by the incredible North American F-86 Sabre. For the first time with the Sabre, US jet fighters were manufactured in significant numbers abroad as well as being exported from America.

The next big hit would come less than a decade later with Lockheed's F-104 Starfighter – a type largely unwanted by the USAF which became a huge sales success abroad, being picked up by the majority of NATO countries albeit under somewhat… controversial circumstances in some cases.

The first truly modern US jet fighter to be widely exported was the McDonnell Douglas F-4 Phantom II. Designed from the outset with flexibility in mind, it proved to be a good fit for many air forces and remains in service today thanks to a host of upgrade programmes. More successful even than the venerable Phantom was Northrop's relatively simple and lightweight – but also highly dependable and long-lasting – F-5 Freedom Fighter/Tiger II. The F-5 has served with more air forces than any other US fighter and continues to serve despite production lines having long since closed.

Then during the 1970s, in the wake of the Vietnam War, came an incredible trio of fighters – the McDonnell Douglas F-15, General Dynamics F-16 and Northrop-developed/McDonnell Douglas-built F/A-18. Together these superbly capable multirole aircraft continue to provide the backbone of US-aligned nations' air forces all over the world. The F-35, it seems, is picking up where they left off – although their stories are by no means over, with new examples of all three (allowing that the F/A-18E/F continues the lineage of the earlier models) still rolling off production lines.

This publication cannot, of course, provide every detail about every exported airframe ever supplied by the US; the myriad upgrade programmes rolled out over the years, the many political twists and turns of military hardware for export, and the detailed operational histories of many thousands of aircraft are beyond the scope of the text contained herein. I have instead attempted to provide the bare essentials for each type and its operators to accompany the beautiful artworks of renowned aviation illustrator JP Vieira – which are the main feature of this book. I hope you enjoy marvelling at the incredible variety of designs as much as I have.

Dan Sharp

ABOUT THE ARTIST

JP Vieira is an illustrator producing military history and aviation-themed artwork.

He is entirely self-taught and aims to constantly improve both the technical and digital methods. His attention to detail and constant pursuit of improvement makes his artworks both accurate and artistically pleasing.

JP is a published artist, collaborating with several authors, editors and publishers.

GENERAL DYNAMICS F-16 FIGHTING FALCON

An F-16 in Turkish service.

US JET FIGHTERS IN FOREIGN SERVICE

CONTENTS

006 | LOCKHEED F-80C SHOOTING STAR ▶

◀ **010** | MCDONNELL F2H BANSHEE

012 | GRUMMAN F9F PANTHER AND COUGAR ▶

◀ **014** | REPUBLIC F-84 THUNDERJET

022 | REPUBLIC F-84F THUNDERSTREAK ▶

◀ **028** | NORTH AMERICAN F-86 SABRE

038 | NORTH AMERICAN F-86D AND F-86K ▶

◀ **044** | NORTH AMERICAN F-100 SUPER SABRE

048 | MCDONNELL F-101 VOODOO ▶

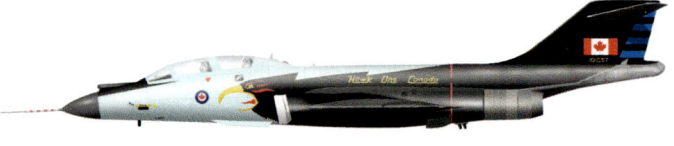

050 | CONVAIR F-102 DELTA DAGGER

052 | LOCKHEED F-104 STARFIGHTER

060 | VOUGHT F-8 CRUSADER

062 | MCDONNELL DOUGLAS F-4 PHANTOM II

072 | NORTHOP F-5 TIGER II

084 | GRUMMAN F-14 TOMCAT

086 | MCDONNELL DOUGLAS F-15 EAGLE

094 | GENERAL DYNAMICS F-16 FIGHTING FALCON

112 | MCDONNELL DOUGLAS F/A-18A, B, C, D HORNET

120 | BOEING F/A-18E, F SUPER HORNET

122 | LOCKHEED F-35 LIGHTNING II

All illustrations:
JP VIEIRA

Design:
DRUCK MEDIA PVT. LTD.

Publisher:
STEVE O'HARA

Production editor:
DAN SHARP

Published by:
MORTONS MEDIA GROUP LTD, MEDIA CENTRE, MORTON WAY, HORNCASTLE, LINCOLNSHIRE LN9 6JR

Tel. 01507 529529

Printed by:
WILLIAM GIBBONS AND SONS, WOLVERHAMPTON

ISBN: 978-1-911703-05-1

© 2023 MORTONS MEDIA GROUP LTD. ALL RIGHTS RESERVED. NO PART OF THIS PUBLICATION MAY BE REPRODUCED OR TRANSMITTED IN ANY FORM OR BY ANY MEANS, ELECTRONIC OR MECHANICAL, INCLUDING PHOTOCOPYING, RECORDING, OR ANY INFORMATION STORAGE RETRIEVAL SYSTEM WITHOUT PRIOR PERMISSION IN WRITING FROM THE PUBLISHER.

US JET FIGHTERS IN FOREIGN SERVICE

LOCKHEED
F-80 SHOOTING STAR

The F-80 was the United States' only operational jet when the USAF was formed in 1947 and it remained in American service till 1958. Exports of around 130 surplus F-80C airframes commenced in 1957 – exclusively to six South American nations.

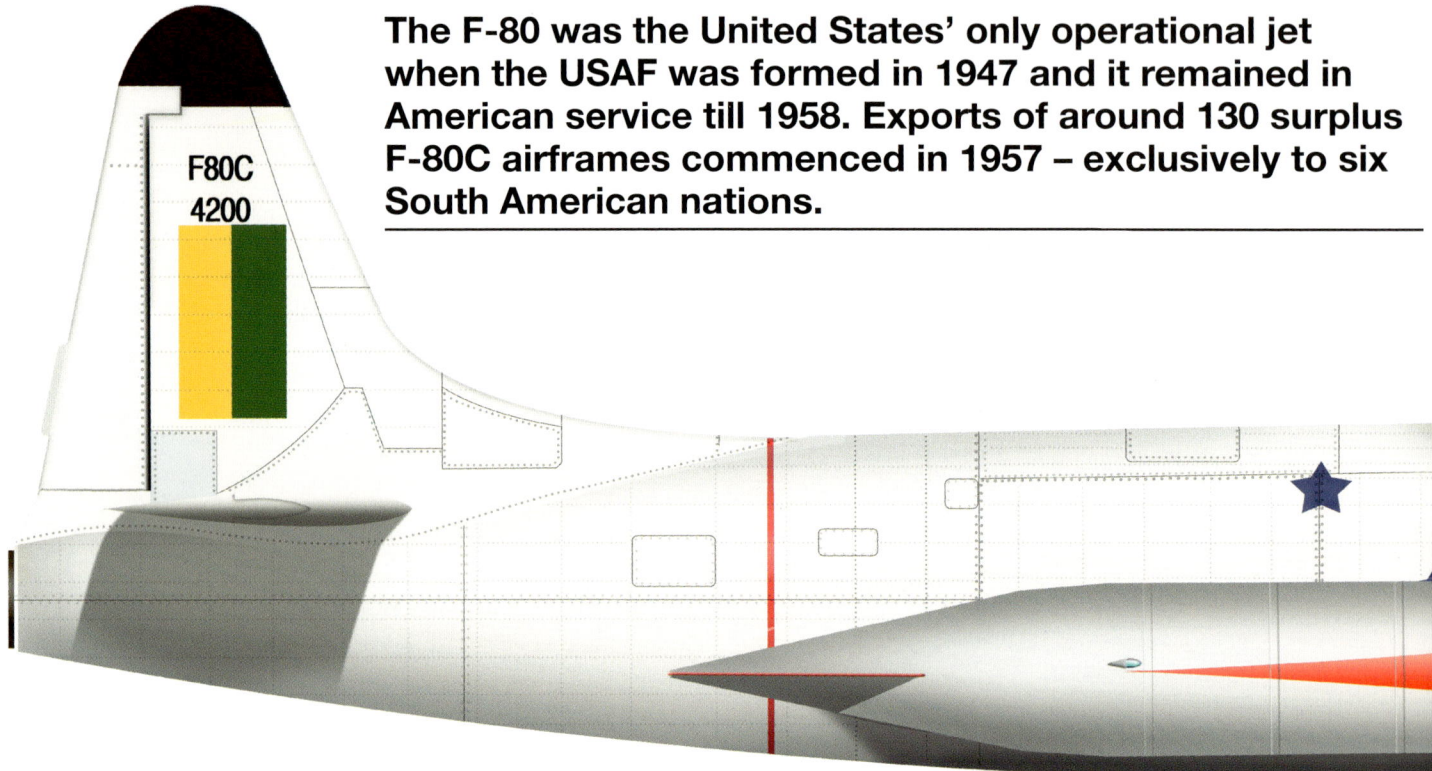

1957-1970

The Lockheed engineering team led by Clarence L 'Kelly' Johnson were given the task of designing a new jet fighter around the British Halford H-1B engine on May 17, 1943. A design proposal was completed the following month and the first prototype was subsequently manufactured within just 143 days.

The airframe was delivered to Muroc Army Airfield in Southern California, today known as Edwards Air Force Base, on November 16 but the first engine delivered from Britain was damaged during a ground test the following day when the aircraft's intake ducts collapsed. A replacement was urgently sought and the first XP-80, serial 44-83020, first flew on January 8, 1944. It was fitted with the last working H-1 then in existence – which had had to be removed from the prototype of the de Havilland Vampire before it could be shipped to the US.

Tests showed that the aircraft, nicknamed 'Lulu-Belle', could reach a top speed of 502mph at 20,480ft. The second and third prototypes, both given the new XP-80A designation, were powered by an American development of British jet pioneer Frank Whittle's engine designs – the General Electric I-40, later built by Allison as the J-33. The first was painted in pearl grey, earning it the nickname 'Grey Ghost', while the second was left unpainted and later became known as the 'Silver Ghost'.

The 'Ghosts' were both used for engine and intake testing and while this was under way some examples of the next batch of 12 aircraft, designated YP-80A, began to enter service in late 1944. A 13th YP-80A was built and modified for photo reconnaissance but was destroyed in a crash in December 1944.

Just two pre-production YP-80A Shooting Stars saw active service during the Second World War, operating briefly from Lesina airfield in Italy with the 1st Fighter Group. Another two were

▼ LOCKHEED F-80C SHOOTING STAR

Lockheed F-80C Shooting Star, 4200, *1º/4º Grupo de Aviação, Força Aérea Brasileira* (1st /4th Aviation Group, Brazilian Air Force), Fortaleza Air Base, Ceará, Brasil, 1960. Brazil operated 33 examples of the Shooting Star, using them for operational conversion. This was the commander's aircraft painted in a special colour scheme.

▼ LOCKHEED F-80C SHOOTING STAR

Lockheed F-80C Shooting Star, J-342, *Grupo n.º 12, Fuerza Aérea de Chile,* (Group No. 12, Chilean Air Force), Chabunco Air Base, Punta Arenas, Chile, 1967.
In the latter years of its operational career, Chilean Shooting Stars were painted in camouflage colours; several were deployed to southern Chile, in the midst of border tensions with Argentina.

US JET FIGHTERS IN FOREIGN SERVICE — Lockheed F-80 Shooting Star

stationed at RAF Burtonwood in Cheshire for demonstration and test flying only.

Powered by a single J-33-GE-9 jet engine mounted centrally in its fuselage, the production model Shooting Star was aerodynamically clean and was therefore able to reach an impressive 536mph in level flight at 5000ft – though only when fully painted and without wingtip fuel tanks.

In natural metal finish and with range-extending wingtip tanks, performance tests showed top speed to be just over 500mph – placing it behind most of its jet-powered contemporaries. Nevertheless, 344 P-80As were ordered in February 1945. The P-80B had an improved J-33 engine and, for the first time, an ejection seat – which was then retrofitted to the remaining P-80As. The P-80C had the largest production run, with 798 examples built. In addition, 129 P-80As were upgraded to P-80C spec for a total of 927 P-80Cs. The type was then redesignated F-80C in 1947.

Both the F-80C and RF-80 photo reconnaissance version saw action during the Korean War – which quickly demonstrated that the F-80 had already been superseded by the next generation of Soviet jet fighters. A total of 277 F-80s were lost on active service during the Korean War – including 113 shot down by anti-aircraft fire and 14 in air-to-air combat.

The F-80C continued in USAF and Air National Guard service into the mid-1950s and during January and February 1957, the American government decided to prop up the ailing Ecuadorian government by supplying a squadron of 13 F-80Cs under its Military Assistance Program. These would be supplemented by three more and eventually six surviving examples were returned to the US in 1965.

The same programme was rolled out to other South American nations

the following year, providing F-80Cs as replacements for aging P-47s that had been supplied some years earlier. Chile received 18, Colombia got 16 and Uruguay got 14. Brazil got the most – a total of 33 – and Peru received 16. The latter had actually ordered four brand new F-80Cs in 1947 but these had never been delivered. Only Peru's F-80s would see anything close to action, with one being used to persuade a rebellious garrison to surrender.

The last operational exported F-80s were retired by Uruguay in 1975 to 1970

▼ LOCKHEED F-80C SHOOTING STAR

Lockheed F-80C Shooting Star, FT-642, *Escuadron de Combate 2112, Ala de Combate No. 21, Fuerza Aérea Ecuatoriana* (2112 Fighter Squadron, N.° 21 Combat Wing, Ecuadorian Air Force), Taura Air Base, Guayas Province, Ecuador, 1960.
Ecuadorian F-80s operated for around seven years, before being returned to the USA in 1965.

▼ LOCKHEED F-80C SHOOTING STAR

Lockheed F-80C Shooting Star, 221, *Grupo de Aviacion n.° 2 (Caza), Fuerza Aérea Uruguaya*, (Aviation Group N.° 2 (Fighter), Uruguayan Air Force), Tte. 2° Mario W. Parallada Air Base, Durazno, Uruguay, 1958.
Shooting Stars were the first combat jets operated by the *Fuerza Aérea Uruguaya* and they were used until the late 1960s.

US JET FIGHTERS IN FOREIGN SERVICE

MCDONNELL F2H BANSHEE

The basic McDonnell F2H Banshee was essentially a bigger Phantom with better engines but the variant exported, the F2H-3, was larger still with a longer fuselage, new wings and a new tail. Canada was the type's sole overseas operator.

1955-1962

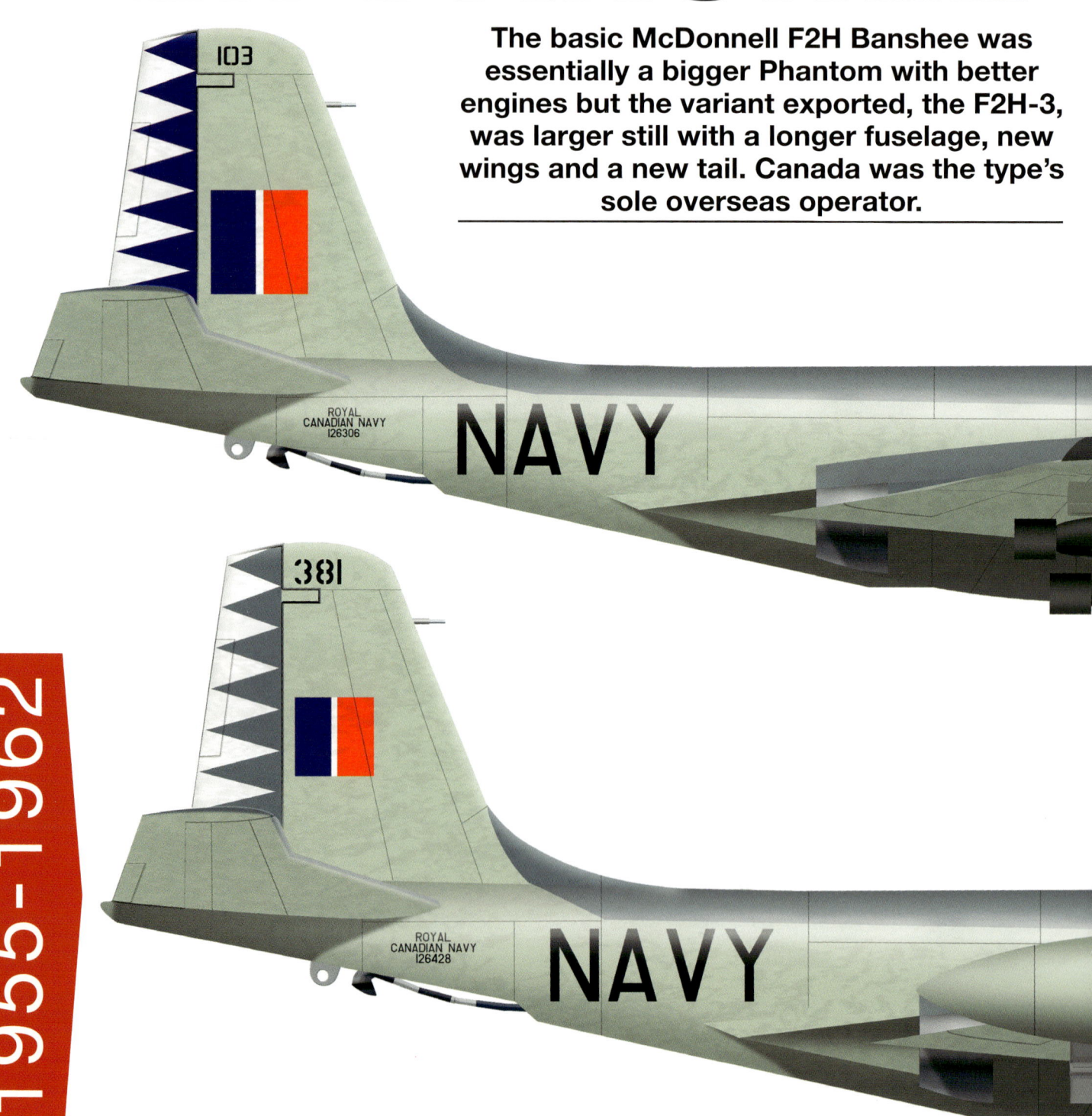

McDonnell received a contract to build three prototypes of its scaled-up Phantom design on March 2, 1945. These would each be powered by a pair of 3000lb thrust Westinghouse J34-WE-22 turbojets and armed with four 20mm cannon clustered together in the nose. The foldable wings were straight and each had four external hardpoints, allowing the Banshee to carry bombs or rockets weighing up to a total of 1540lb.

The first prototype XF2H-1 flew on January 11, 1947, and an order for 56 production model F2H-1s followed on May 29. An order for the improved F2H-2 was placed even before the first F2H-1s had been delivered. This second variant was the most-produced Banshee with 364 examples rolling off McDonnell's St Louis production line.

The F2H-2 was powered by J34-WE-34s, producing 3250lb of thrust, had an ejection seat and could be fitted with wingtip fuel tanks for improved range. Fourteen became F2H-2N night and all-weather interceptors with the addition of the AN/APS-19A radar in a new lengthened plastic nose section, the cannon being shifted further aft.

With the success of the F2H-2N, McDonnell now proposed a thoroughly revised Banshee as a purpose-built all-weather interceptor – the F2H-3. This had the same engines of its predecessor but inside a 7ft longer fuselage with enlarged wings and a completely redesigned tail unit. It maintained some parts commonality but was practically a new aircraft. The fuselage could house two new fuel tanks, which meant that tip tanks were no longer always necessary, and inside the nose was an AN/APQ-41 radar. The US Navy ordered 250 examples.

Meanwhile, the Royal Canadian Navy wanted new F2H-3s to replace its aging piston-engined Hawker Sea Fury F.B.11s, but by the time the necessary funds were in place McDonnell's Banshee production line had already been dismantled. The Canadians therefore purchased 39 ex-US Navy F2H-3s.

The first of these were collected by RCN pilots from NAS Quonset Point, Rhode Island, on November 30, 1955, and flown to Shearwater, Nova Scotia.

Three RCN Banshee units were established – VF-870, VF-871 and VX-10, a test squadron – and they finally received the last of their 'new' Banshees on June 16, 1958. While the US Navy retired its Banshees from front line service in 1959, the Canadians continued to operate them for three more years, sometimes aboard the carrier HMCS *Bonaventure*. During that time most were wrecked as a result of accidents so that few remained when VF-870, the last of the three Banshee units, was disbanded in 1962.

▼ MCDONNELL F2H-3 BANSHEE

McDonnell F2H-3 Banshee, 103, 871 Naval Air Squadron (VF-871), Royal Canadian Navy/Marine Royale Canadienne, HMCS *Bonaventure*, 1956. Canadian Banshees were shore-based at Nova Scotia but also saw deployment aboard the HMCS *Bonaventure* aircraft carrier.

▼ MCDONNELL F2H-3 BANSHEE

McDonnell F2H-3 Banshee, 381, 870 Naval Air Squadron (VF-870), Royal Canadian Navy/Marine Royale Canadienne, Royal Canadian Naval Air Station Shearwater, Nova Scotia, 1958. Canadian Banshees could carry the AIM-9B Sidewinder air-to-air missile, supplementing their guns for aerial combat.

US JET FIGHTERS IN FOREIGN SERVICE

GRUMMAN F9F PANTHER/COUGAR

Argentina was the only foreign operator of Grumman's capable F9F series jet fighters – the straight-winged Panther and its swept wing successor, the Cougar.

The US Navy chose a single-engine all-metal straight-winged jet fighter design proposed by Grumman for development as the XF9F-2 and XF9F-3 in early 1947. The former would have Pratt & Whitney's 5000lb thrust J42-P-6, while the latter had Allison's 4600lb thrust J33-A-8.

The first XF9F-2 flew on November 24, 1947, but the first XF9F-3 would not follow till August 16, 1948. The Navy ordered 47 F9F-2s and 54 F9F-3s but when they began to arrive in 1949 found that the F9F-2 offered superior performance while the F9F-3s were plagued with technical problems. Most F9F-3s were later retrofitted with the J42.

Named 'Panther' by Grumman, the new type had a tricycle undercarriage, foldable wings with flaps on both leading and trailing edges, and straight tailplanes high up on the tail fin. The intakes for the centrally-mounted engine were in the wingroots. Total fuel capacity with tip tanks was 923 gallons. The pressurised cockpit featured an ejection seat and

1958-1971

▼ GRUMMAN F9F-8T COUGAR

Grumman F9F-8T Cougar, 3-A-152, *1.ª Escuadrilla Aeronaval de Ataque, Comando de Aviación Naval, Armada Argentina* (1st aero-naval attack flight, Naval Air Command, Argentinian Navy), Punta Indio naval air base, Buenos Aires, Argentina, 1963
The Argentinian Navy used the F9F-8T for conversion training; this was their first aircraft with in-flight refuelling capability.

armament was four 20mm M3 cannon in the lower nose. Three, and later four, hardpoints under each wing could be used for bombs or rockets.

Encouraged by the success of the F9F-2, Grumman began working up plans to give the Panther 35° swept back wings, with 40% more surface area, and swept tailplanes. The company proposed this change to the Navy and on March 2, 1951, was given a contract to build three examples under the designation XF9F-6.

Work progressed quickly and the first prototype flew just over six months later on September 20. Tests showed that the modified Panther was now 50mph faster so an order was placed and deliveries of the F9F-6 and F9F-7 Cougar commenced. Further development led to the F9F-8 with redesigned wings, a stretched fuselage and a J48-P-8A engine which had 8500lb of thrust with water injection.

The first F9F-8 flew on December 18, 1953, and deliveries began on February 29, 1954, continuing up to March 22, 1957. The F9F-8T trainer variant, with another fuselage extension to accommodate a second cockpit behind the original one, was also built. Internal fuel tank capacity was reduced and two of the four cannon were removed.

Argentina bought 24 refurbished F9F-2s for the Argentine Navy in 1958 and they were operated by the 1a Escuadrilla Aeronaval de Ataque from land bases since the catapult on the light carrier ARA *Independencia* was deemed too weak to launch them. Two F9F-8Ts were obtained by Argentina in 1962.

During a military coup in 1963, which was supported by the Argentine Navy, F-86 Sabres, Gloster Meteors and MS.760s of the Argentine Navy attacked a naval base and destroyed four Panthers on the ground. Surviving Panthers were then used to fly patrols during a border dispute between Chile and Argentina in 1965.

The fleet was finally grounded due to lack of spares in 1969, with the two Cougars remaining in service until 1971.

▼ GRUMMAN F9F-2 PANTHER

Grumman F9F-2 Panther, 3-A-110, *1.ª Escuadrilla Aeronaval de Ataque, Comando de Aviación Naval, Armada Argentina* **(1st aero-naval attack flight, Naval Air Command, Argentinian Navy), Punta Indio naval air base, Buenos Aires, Argentina, 1960**
Although Panthers could land on the aircraft carrier ARA *Independencia*, the catapult was not powerful enough to launch them so they operated from land bases.

US JET FIGHTERS IN FOREIGN SERVICE

REPUBLIC F-84 THUNDERJET

The Thunderjet's F-84G variant was built in huge numbers and more than half of those bought for the USAF were sent instead to around a dozen foreign nations via the Mutual Defense Assistance Program.

1951-1974

The team at Republic Aviation, overseen by company chief designer Alexander Kartveli, began work on their first jet fighter project in 1944 under the designation AP-23. It was hoped at first that time could be saved by utilising as much of the company's successful P-47 Thunderbolt airframe as possible, but this was soon found to be impractical and a whole new aircraft was developed instead. The fighter was to be powered by a single General Electric TG-180 turbojet – the production version of which would be known as the J35-GE-7.

After a year of development work, Republic was finally awarded a contract for three prototypes and a fourth airframe for static testing on November 11, 1945, under the designation XP-84.

The order was increased to 25 pre-production service evaluation aircraft and 75 production models on January 4, 1946, the balance later being shifted to 15 and 85 respectively. The first prototype commenced flight testing on February 28, 1946, at Muroc Air Base in

▼ REPUBLIC F-84G THUNDERJET

Republic F-84G Thunderjet, A-769, *730 Eskadrille, Flyvevåbnet*, (730 Squadron, Royal Danish Air Force), Skrydstrup air base, Sønderjylland, Denmark, 1959.
Denmark operated more than 200 Thunderjets; some were later on painted in camouflage colours.

▼ REPUBLIC F-84G THUNDERJET

Republic F-84G Thunderjet, 51-18, *Pattuglia Acrobatica Tigri Bianchi, 51ª Aerobrigata, Aeronautica Militare* (White Tigers aerobatic Team, 51st Air Brigade, Italian Air Force), Istrana air base, Veneto, Italy, 1956.
The aerobatic team performed for only a year – from 1955 to 1956.

US JET FIGHTERS IN FOREIGN SERVICE — Republic F-84 Thunderjet

▼ REPUBLIC F-84G THUNDERJET

Republic F-84G Thunderjet, Golden Crown aerobatic team, Imperial Iranian Air Force, Mehrabad International Airport, Tehran, Iran, 1960.
The IIAF's Golden Crown team would operate from 1958 until 1979, with the Thunderjets being the first aircraft flown.

▼ REPUBLIC F-84G THUNDERJET

Republic F-84G Thunderjet, FS-746, *337 Mira, Polemikí Aeroporía* (337 Squadron, Hellenic Air Force), Larissa air base, Thessaloniki, Greece, 1956.
337 Mira was the first HAF Thunderjet squadron; a demonstration team with four aircraft was formed within the squadron.

▼ REPUBLIC F-84E THUNDERJET

Republic F-84E Thunderjet, TP-6, 306 Squadron, *Koninklijke Luchtmacht* (Royal Netherlands Air Force), Volkel air base, North Brabant, Netherlands, 1954.
306 Squadron used Thunderjets for reconnaissance, with modified wingtip tanks containing cameras; this aircraft is equipped with JATO bottles to assist in take-offs.

California, now Edwards AFB, with Major William A Lein at the controls. It had a very simple layout – a nose intake for the centrally positioned J35, straight wings, a conventional tail, hydraulically actuated tricycle undercarriage and a sliding bubble canopy for the cockpit.

The fuselage was too slender to house both the engine and its fuel tanks, so the wings were made thick enough to carry the aircraft's fuel supply. A hydraulic dive brake was installed beneath the fuselage in the middle.

The first flight of the second XP-84 was in August 1946 and the aircraft set a new American national air speed record on September 7 at 611mph – the world record of 615.78mph being held by the British Gloster Meteor F.4.

General Electric was struggling with the J35-GE-7 so the third prototype, designated XP-84A, was powered by the Allison-built J35-A-15 and this was used for the 15 pre-production XP-84As. These also had four nose-mounted 0.50-calibre M2 Browning machine guns and another one in each of their wing roots. They could also be equipped with two 230 gallon wingtip tanks. The full production version was the P-84B, which had a slightly improved engine, M3 machine guns and an ejection seat – although the latter was never approved for use.

The type entered service with the 14th Fighter Group in December 1947 and it was quickly found to have serious problems. It had the power to go faster than Mach 0.8 but above that speed at low altitude it could suffer from control reversal and sudden violent pitch-up which could cause its wings to snap off. Above 15,000ft it was possible to fly faster but only with severe buffeting. Nevertheless, most YP-84As would later be brought up to P-84B standard and a total of 226 P-84Bs were delivered to the air force between August 1947 and February 1948.

The entire P-84B fleet was grounded on May 24, 1948, following a series of structural failures. The P-84C, similar to the 'B' but with an improved electrical system and engine, began production in May 1948 and the type was then redesignated F-84 on June 11, 1948. Both the F-84B and F-84C suffered from a host of mechanical issues so Republic produced the F-84D, embodying various quick fixes ahead of the fully redesigned F-84E.

The F-84D had the more powerful Allison J35-A-17D engine, mechanically rather than hydraulically actuated landing gear, stronger wings, thicker aluminium skin, wingtip tank fins to prevent flexing, a winterised fuel system, a usable ejection seat and a quick-release cockpit canopy. A total of 154 F-84Ds were delivered between November 1948 and April 1949. The first F-84E, first flown on May 18, 1949, had further wing reinforcement, a radar gunsight, provision for rocket-assisted take-off, a 12in longer fuselage forward of the wings and a 3in extension aft of the wings to expand the avionics bay. It also had rocket racks

US JET FIGHTERS IN FOREIGN SERVICE — Republic F-84 Thunderjet

▼ REPUBLIC F-84G THUNDERJET
Republic F-84G Thunderjet, MU-Z, *338 Skvadron*, Luftforsvaret (338 Squadron, Royal Norwegian Air Force), Ørland main air station, Trøndelag, Norway, 1959.
Receiving more than 200 examples of the F-84, the Thunderjet was the first US jet of the RnoAF and the one, so far, operated in the largest numbers.

▼ REPUBLIC F-84G THUNDERJET
Republic F-84G Thunderjet, 1214/923, 12th Squadron, 1st Wing, Royal Thai Air Force, Don Muang air base, Bangkok, Thailand, 1957.
The RTAF operated Thunderjets until the 1960s when they were replaced by F-86 Sabres.

which folded flat against the wing once their rockets had been fired.

Between May 1949 and July 1951, 843 F-84Es were built and many saw action during the Korean War, though they were soon switched to the ground-attack role having proven inferior to the MiG-15 in air-to-air combat.

The last straight-wing variant was the F-84G, which entered service in 1951 ahead of the introduction of the swept-wing F-84F. It featured a more powerful J35-A-29, a framed canopy, refuelling boom, autopilot, instrument landing system and option to carry a single Mark 7 nuclear bomb. The F-84G proved, at last, to be both capable and reliable – with the USAF eventually ordering 3025 examples. However, with still more powerful and capable types now entering production, a total of 1936 were sent directly from the US to foreign nations – starting with Belgium, France, the Netherlands and Norway in 1951, followed by Denmark, Greece, Italy and Turkey in 1952. Portugal, Taiwan and Yugoslavia also received F-84Gs in 1953. Thailand joined the F-84G club in 1956 and finally Iran received the type in 1957.

Some of these aircraft would end up serving with more than one air force – since when the type was retired by one nation, surviving airframes with remaining

▼ REPUBLIC F-84G THUNDERJET
Republic F-84G Thunderjet, 117, 23rd Squadron, 4th Wing, Republic of China Air Force, Chiayi Air Base, Taiwan, 1956.
Taiwanese Thunderjets clashed with PRC's MiGs, claiming several kills. The faint former USAF buzz number (FS-374) can still be seen beneath the new markings of this F-84.

US JET FIGHTERS IN FOREIGN SERVICE

US JET FIGHTERS IN FOREIGN SERVICE — Republic F-84 Thunderjet

hours were typically moved on to another which continued to operate it. Hence, the 'total operated' figures below add up to considerably more than 1936.

The first four foreign operators would also receive a handful of F-84Es, with Belgium having 21 F-84Es and 192 F-84Gs, France receiving 140 F-84Es and F-84Gs, the Netherlands getting 21 F-84Es and 179 F-84Gs, and Norway receiving six F-84Es and 200 F-84Gs. While the first three would all retire their Es and Gs in 1957, Norway would continue to operate the F-84G into 1960.

Denmark would be the last country to get any F-84Es, receiving six, plus 240 F-84Gs, and it operated them until January 1962. After Denmark, only the F-84G would be sent abroad. Greece got 200, flying them till June 1960; Italy received 254, retiring them in 1957, and Turkey got a remarkable 479 which it continued to fly until June 1966.

Portugal received 125 which it flew until 1973; Taiwan (Republic of China) got 246 which it retired in 1964 and Yugoslavia got 219 which remained in service until 1974. Thailand's modest fleet of 31 F-84Gs would last just 10 years – up to 1966 – and Iran had a similarly small contingent of the type which it retired in February 1965.

The only foreign-operated F-84Gs to see actual combat were those of the Republic of China Air Force. They were

▼ REPUBLIC F-84G THUNDERJET
Republic F-84G Thunderjet, 5168, *Esquadra 93 'Magníficos', Força Aérea Portuguesa* (93 Squadron, Portuguese Air Force), air base 9, Luanda, Angola, 1966. Portugal use its F-84s during the colonial wars, both in Angola and Mozambique, from 1961 to 1974.

involved in several clashes with People's Liberation Army Air Force MiG-15s and MiG-17s during the mid-to-late 1950s. During one battle on July 21, 1956, ROCAF pilot Heinz Ouyamg was credited with shooting down two MiG-17s in his F-84G – a feat for which he received Taiwan's second highest military award, the Order of Blue Sky and White Sun.

Two years later, during the Second Taiwan Strait Crisis, ROCAF F-86s and F-84Gs participated in around 25 engagements and were credited with destroying 32 enemy aircraft for a loss of just two F-86s and two F-84Gs. Most of the ROCAF kills were, however, a result of the F-86s having recently been equipped with Sidewinder missiles.

The F-84G may not have set the world alight with its stellar performance, but it was a dependable strike fighter and in the right hands could be a very capable combat aircraft.

▼ REPUBLIC F-84G THUNDERJET

Republic F-84G Thunderjet, 595, *172. lovačkobombarderski avijacijski puk -školski, Ratno Vazduhoplovstvo i Protivvazdušna Odbrana* (172nd Fighter-Bomber Aviation Regiment- Training, Yugoslav Air Force and Air Defence), Zemunik air base, Yugoslavia, 1966.
The Yugoslav Air force was a recipient of early US fighter jets, starting with more than 200 Thunderjets; they were operated until the mid-1970s.

US JET FIGHTERS IN FOREIGN SERVICE

REPUBLIC F-84F THUNDERSTREAK

Swept-wing F-84Fs replaced the straight-wing F-84Gs of four European NATO states during the mid-to-late 1950s as well as equipping the newly re-formed Luftwaffe in West Germany. Surviving airframes would then be passed on to two further foreign operators in the 1960s.

1955–LATE 1970s

▼ **REPUBLIC F-84F THUNDERSTREAK**

Republic F-84F Thunderstreak, 1-ET/725, *Escadron de Chasse 2/1 Morvan, Armee de l'Air* (Fighter Squadron 2/3, French Air Force), 113 Air Base, Saint-Dizier, France, 1958.
The French Air Force operated several early US jets alongside domestically developed aircraft such as the Dassault Ouragan and Mystere.

As F-84E production was getting under way in late 1949, Republic began to consider whether swapping its straight wings for swept ones would result in superior performance. The 409th F-84E off the production line was therefore given wings with a 40° sweep-back.

It also had a J35-A-25 engine with 5200lb-ft of thrust – compared to the J35-A-17D of the standard 'E' which developed 5000lb – and its windscreen was redesigned to improve its aerodynamic profile. The aircraft, initially designated YF-96A, made its first flight on June 3, 1950, and achieved a respectable 693mph during one low-altitude test.

Republic received a development contract the following month, with the Air Force requiring that the J35 be swapped for a Curtiss-Wright J65, a licence-built version of the British Armstrong Siddeley Sapphire engine.

Two months later the YF-96A was redesignated YF-84F and received a new nickname – Thunderstreak.

The lone YF-84F was next given a deepened fuselage to accommodate the J65 and flew with it for the first time on February 14, 1951. Two purpose-built YF-84F prototypes followed – one of which featured a solid nose and intakes in its wingroots.

Manufacturing and engine installation difficulties meant that the first production model F-84F did not fly until November 22, 1952, with the F-84G entering production in the meantime as a stopgap. In addition to its new wings, engine and fuselage, the Thunderstreak also differed

▼ REPUBLIC F-84F THUNDERSTREAK

Republic F-84F Thunderstreak, FU-57/ YL-H, 3rd Squadron, 2nd Wing, *Belgische Luchtmacht – Force Aérienne Belge* (Belgian Air Force), Florennes air base, Belgium, 1957. The Belgian Air Force operated the Thunderstreak until the early 1970s being eventually replaced by the Lockheed F-104G and Dassault Mirage 5BA.

US JET FIGHTERS IN FOREIGN SERVICE — Republic F-84F Thunderstreak

▼ REPUBLIC F-84F THUNDERSTREAK

Republic F-84F Thunderstreak, 1-NX/A, *Escadron de Chasse 1/1 Corse, Armee de l'Air* (Fighter Squadron 1/1, French Air Force), Lod Air Base, Israel, 1956.
During the 1956 Suez Canal Crisis, French F-84Fs were painted in 'invasion stripes' and some even carried Israeli Air Force markings, operating from Israeli bases in cooperation with British and Israeli forces.

▼ REPUBLIC F-84F THUNDERSTREAK

Republic F-84F Thunderstreak, DB-113, *Jagdbombergeschwader 32, Luftwaffe*, Lechfeld, Westphalia, West Germany, 1962.
A major operator of the Thunderstreak, the Luftwaffe fielded 450 examples for a period of 10 years.

▼ REPUBLIC F-84F THUNDERSTREAK

Republic F-84F Thunderstreak, 837, *344 Mira, Polemikí Aeroporía* (334 Squadron, Hellenic Air Force), Larissa air base, Thessaloniki, Greece, 1972.
The Greek Air Force retired the last F-84Fs in 1978.

from the Thunderjet in having a one-piece flip-up cockpit canopy; a pair of air brakes on the rear fuselage sides rather than a single brake underneath; powered flight controls; leading edge slats and a spine running down its back.

It retained the F-84E's six Browning M3 machine guns but its swept wings could not take wingtip fuel tanks so of its four underwing pylons the inner two were designed to carry external tanks. The 'F' could also deliver a nuclear bomb using the LABS launch mechanism.

Although the J65-W-1 engine was fitted to the first 275 Thunderstreaks, this was replaced with the J65-W-1A for the next 100. Neither powerplant was very reliable and aircraft fitted with them could not later be re-engined with improved variants – meaning their service lifespan was limited.

After those first 375 F-84Fs, the somewhat more reliable J65-W-3 and J65-B-3 (manufactured by Buick) were introduced, each providing 7220lb-ft of thrust. Even so, the aircraft had a high landing speed and although it was faster than the straight-winged F-84s it did not handle well when approaching its limit. These problems prompted Republic to suspend production in 1954 and reassess the F-84F's design. When production restarted in 1955, the new aircraft were fitted with 'all-moving' tailplanes – preventing a tendency to pitch-up during a high-speed stall.

The whole fleet was grounded in 1955 due to repeated engine failures. Improvements were made but the J65 could not be raised to a satisfactory standard and the USAF began to phase the F-84F out of active service that year – with spare airframes being earmarked for delivery to America's NATO allies under the Mutual Defense Assistance Program.

The first to receive an allocation of these was France in 1955. French Thunderstreaks equipped five squadrons and during the Suez Crisis of October-November 1956 they were used to attack Egyptian airfields. A total of 20 Egyptian Ilyushin Il-28 Beagle bombers were destroyed on the ground and one was shot down, having taken off. Just one F-84F was lost during the operation. The introduction of the Dassault Mirage IIIE in the mid-1960s saw the Thunderstreaks eventually phased out.

Belgium also got its first F-84Fs during 1955s – replacing F-84Gs – eventually phasing out the last examples during the early 1970s.

The Netherlands was next to begin receiving deliveries, though it would have to wait until 1956 to get its first F-84Fs. These were, again, eventually retired following the introduction of the F-104G. The last of the F-84G operators to receive F-84Fs was Italy, which equipped three air brigades with them, starting in 1956. A number of these would be kept on strength, however, till 1972 even though Italy too bought into the initial phase of F-104G production.

The West German Luftwaffe was re-formed in 1956 and the Thunderstreak

US JET FIGHTERS IN FOREIGN SERVICE — Republic F-84F Thunderstreak

▼ REPUBLIC F-84F THUNDERSTREAK

Republic F-84F Thunderstreak, 5-36591, *Pattuglia Acrobatica Getti Tonanti, 5ª Aerobrigata, Aeronautica Militare* (Thunder Jets Aerobatic Team, 5th Air Brigade, Italian Air Force), Rimini Miramare air base, Italy, 1960. Performing at the inauguration of the 1960 Rome Olympics, the team's aircraft had the Olympic logo painted on their fins.

▼ REPUBLIC F-84F THUNDERSTREAK

Republic F-84F Thunderstreak, P-177, 314 Squadron, *Koninklijke Luchtmacht* (Royal Netherlands Air Force), Eindhoven Air Base, North Brabant, Netherlands, 1970. For the farewell flight of the F-84F, the RNLAF had a very special colour scheme: the Hippy Streak!

▼ REPUBLIC F-84F THUNDERSTREAK

Republic F-84Q Thunderstreak, 7044, *113 Filo, Türk Hava Kuvvetleri (113 Squadron,* Turkish Air Force), 1st Main Jet Air Base, Eskişehir, Turkey, 1972.
The F-84Qs were modified ex-Luftwaffe examples supplied to Turkey.

became its first fighter not long thereafter, with the first German fighter-bomber wing, Jagdbombergeschwader 31, officially becoming operational on June 20, 1958. At one point around 450 Thunderstreaks were serving in West Germany, making it the second largest operator of the type after the USAF. The steady introduction of the F-104G, however, would see those aircraft eventually phased out by the mid-1960s.

Finally, during the late 1950s a number of former German, Belgian and Dutch F-84Fs were transferred to Greece and Turkey, with two Turkish F-84Fs shooting down a pair of Iraqi Il-28s which violated Turkish airspace on August 16, 1962. Both nations would phase out their Thunderstreaks during the late 1970s with the introduction of the Northrop F-5.

Overall, 2711 F-84F Thunderstreaks, including prototypes, had been manufactured by the time production ceased in August 1957 and 1301 of these served with foreign nations.

US JET FIGHTERS IN FOREIGN SERVICE

US JET FIGHTERS IN FOREIGN SERVICE

NORTH AMERICAN F-86/CL-13 SABRE

While the Thunderjet and Thunderstreak were exported in considerable numbers, North American's F-86 dramatically surpassed Republic's fighter-bombers. The Sabre and its derivatives would be flown by dozens of air forces around the world.

1950-1993

The US Navy drew up a requirement for a jet-powered shipboard fighter during the autumn of 1944 and North American responded with project NA-134. The underwhelming result was the FJ-1 Fury, a single engine aircraft with a nose intake plus wings, tail surfaces and a cockpit canopy which owed much to those of the P-51 Mustang.

When the USAAF issued a requirement for a medium-range jet fighter in November 1944, the company amended NA-134 to include a slimmer fuselage and revised wings. It was to have six 0.5in machine guns in its nose – three on either side of the intake – and this became the NA-140. North American received a contract for three prototypes under the designation XP-86 in May 1945.

With the end of the Second World War in Europe that month, like every other American defence contractor North American acquired German aerodynamics test data and research materials. This included details of swept wing studies and as such a swept-wing for the XP-86 was discussed in June 1945. After wind tunnel testing the design's original straight wing was scrapped and work began on a new swept wing. A final wing form had been chosen by October 1946 and on December 20 North American received a contract for 33 full production model P-86As and 190 P-86Bs.

The P-86B, based on North American model NA-152, had larger mainwheels and bigger brakes than the 'A', a correspondingly wider fuselage, increased tail area, improved internal fuel capacity, gun heating and a canopy ejection system – but it was cancelled in September 1947 when it was determined that high-pressure tyres made bigger wheels unnecessary. Another 188 P-86As were ordered instead.

Less than a month later, on October 1, 1947, the first prototype XP-86 made its flight debut. Testing revealed a number of issues – its Allison J35 engine was somewhat underpowered, its elevator assembly made it unstable at high speeds and there were landing gear problems. Nevertheless, after two months of company tests, the first XP-86 was handed over to the USAF – as it now was – for the second phase of its test regime. This lasted just six days, after which the Air Force's test pilot Major Ken Chilstrom declared that the USAF now had the world's best jet fighter. North American then received a production contract for a further 225 P-86As.

Also in December 1947, the Air Force ordered two prototypes of the NA-157, designated P-86C. This was to be a long-range fighter with a solid nose and side intakes but it proved so different from the original series that it was later redesignated YF-93A.

The P-86A differed from the three prototypes in having the much more powerful General Electric J47 engine and a redesigned nosewheel door to cure the earlier undercarriage issue. The openings for the aircraft's six cannon on the sides of the nose were covered with panels which opened automatically when they were fired and closed again afterwards. Numerous internal changes refined the cockpit design and systems layout.

In the air, the P-86A's official top speed was 585mph – making it more than 70mph faster than the prototypes. From June 1, 1948, the P-86A became the F-86A. The first USAF deliveries were made on February 14, 1949.

At around the same time, the USAF issued a requirement for a new radar-equipped interceptor and on March 28, 1949, North American put forward

▼ CAC SABRE
CAC Sabre, A94-359, 76 Squadron, Royal Australian Air Force, Williamtown Air Base, Australia, 1965.
76 Squadron operated the CAC Sabre from 1961 onwards and formed an aerobatic team (the Black Panthers) in 1965.

▼ NORTH-AMERICAN F-86F SABRE
North-American F-86F Sabre, C-120, II Brigada Aérea, Fuerza Aérea Argentina (2nd Air Brigade, Argentinian Air Force), General Urquiza air base, Paraná, Argentina, 1972.
Received in 1960, Argentinian Sabres were put on alert during the Falklands War, before being finally retired in 1986; in the final part of its career, the aircraft left their natural metal finish and were painted in camouflage colours.

▲ CANADAIR CL-13 SABRE MK.4
Canadair CL-13 Sabre Mk.4, XB956/T, 112 Squadron, Royal Air Force, Bruggen Air Base, West Germany, 1955.
112 Squadron was deployed in West Germany and operated the Sabre for two years.

US JET FIGHTERS IN FOREIGN SERVICE — North American F-86/CL-13 Sabre

▼ CANADAIR CL-13 SABRE MK.5
Canadair CL-13 Sabre Mk.5, 21651, Golden Hawks demonstration team, Royal Canadian Air Force/Aviation Royale Canadienne, Chatam air base, New Brunswick, Canada, 1962. From 1959 until 1964, the Golden Hawks flew their appropriately painted Sabres at many air shows.

▼ NORTH AMERICAN F-86F SABRE
North American F-86F Sabre, 276, Imperial Ethiopian Air Force, Debre Zeit air base, Ethiopia, 1964.
Ethiopia operated a squadron of Sabres from 1960; the aircraft would be supplemented by Northrop F-5As in 1966.

project NA-164. This single seater, which became the F-86D, is discussed separately in Chapter 7.

Even as the F-86A was entering service, North American was investigating ways of preventing the loss of elevator control at high speeds identified during the XP-86 flight tests. An extension of the trailing edge had alleviated the problem sufficiently for the F-86A but had not cured it entirely.

The answer proved to be an 'all-moving tail' where the entire tailplane assembly operated as a moveable control surface. Work on designing an F-86 with this feature began on November 15, 1949, with the designation NA-170. The USAF then awarded North American a contract for 111 examples of the aircraft as the F-86E on January 17, 1950.

The F-86E also included a new hydraulic system which prevented external loads on the control surfaces from being transferred to the pilot's control stick. Instead, preloaded springs were used to provide the pilot with a simulation of 'feel' when moving the stick. The first 'E' took to the air on September 23, 1950, and the first deliveries to the air force took place in February 1951. A total of 456 were made.

The definitive F-86 day fighter appeared in the form of the F-86F – with work on the new version commencing on July 31, 1950, as the NA-172. The main difference between the 'E' and the 'F' was the installation of the much improved J47-GE-27 engine. It was also fitted with a gunsight-mounted gun camera and had provision to carry AN-M10 chemical tanks on its external pylons.

CANADIAN SABRES

There are three separate strands when it comes to F-86 exports – ex-USAF F-86F Sabres that were sent abroad, licence-made Commonwealth Aircraft Corporation Sabres which saw limited exports and licence-made Canadian Canadair CL-13 Sabres, many of which were also exported around the world. Canada signed a deal to build its own Sabres in 1949 and the Canadian government ordered 100 examples from Montreal-based Canadair in August of that year. It was soon decided these would to be based on the latest variant at the time, the F-86E, but the first one, the Sabre Mk.1, was essentially an F-86A assembled mostly from American-supplied parts. It made its maiden flight on August 9, 1950.

The Sabre Mk.2 was basically an F-86E and the first one flew on January 31, 1951, with another 349 being made between January 1951 and August 1952. 410 Squadron of the RCAF received its first Mk.2 on April 10, 1951. Canada soon had so many Sabres that the USAF, short of aircraft during the Korean War, bought 60 of them itself in February 1952 and delivered them to operational units on the front line.

The 100th Sabre Mk.2 off the production line was fitted with the Canadian-made Avro Orenda engine and became the sole Sabre Mk.3 in the process. It had been hoped that the Orenda could be fitted to the Sabre Mk.4 but the project wasn't ready in time, so the type's only improvements were cockpit air conditioning, a new compass and a new canopy release mechanism. Even so, 438 Mk.4s were built – many later retrofitted with North American's '6-3' wing modification, which eliminated the leading edge slats but added a leading edge extension of 6in at the root changing to 3in at the tip (hence the name). Wing fences were also included.

Most of the Mk.4s built would end up being sent to serve with the RAF in Britain as an interim type until the Supermarine Swift and Hawker Hunter became available. The RAF received three Mk.2s in October 1952 then Mk.4

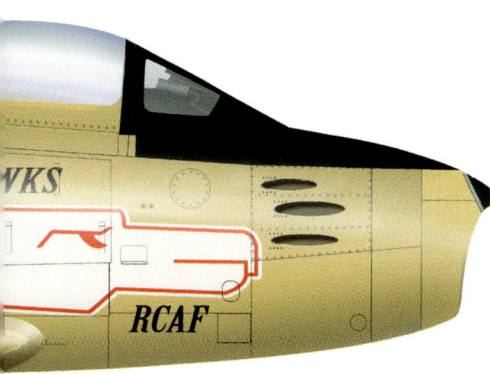

▼ NORTH AMERICAN F-86F SABRE

North American F-86F Sabre, FAB-656, *Escuadrón de Caza 320, Grupo Aereo de Caza 32, Fuerza Aérea Boliviana* (320 Fighter Squadron, 32 Fighter Group, Bolivian Air Force), El Trompillo air base, Santa Cruz, Bolivia, 1991.
Bolivia was the last military operator of the Sabre, retiring the aircraft in the early 1990s.

US JET FIGHTERS IN FOREIGN SERVICE — North American F-86/CL-13 Sabre

deliveries commenced two months later. Altogether, the RAF received a total of 430 Mk.2s and Mk.4s – giving them the British designations Sabre F.2 and F.4 respectively. Ten squadrons in Germany and another two in Britain would eventually operate the type but the Germany-based units had all replaced their Sabres with Hunters by June 1956. Surviving aircraft were then passed on to Italy (180 F.4s) and Yugoslavia (121 F.4s). From 1956 to 1958, all other British Sabres were sent to the USAF, which designated them F-86E(M), with the 'M' standing for 'Modified' since they were not exactly the same as standard American F-86Es.

In Italian service, the Mk.4s also became F-86E(M)s, with deliveries completed by the end of 1957. The Frecce Tricolori display team flew Sabres in 1961 and five Italian Sabres were involved in UN peacekeeping duties in the Belgian Congo in 1963. The type was phased out of Italian service from 1963 to March 1965.

Yugoslavia's Jugoslovensko Ratno Vazduhoplovstvo (JRV) got its ex-RAF F.4s between 1958 and 1959. During this period, one shot down a Hungarian MiG-15 in a border incident. Eight of these F-86E(M)s were eventually transferred on to Honduras in 1967 where they were involved in harassment operations over El Salvador in 1969. These aircraft would end up serving into the 1980s before eventually being replaced with Dassault Super Mystères.

Canadair continued to upgrade the CL-13, creating the Mk.5 and Mk.6 variants. With these becoming available, Canada began to sell off its older Sabres. Greece bought 104, starting in July 1954, and operated them until the early 1960s. Turkey meanwhile bought 105 – operating them during a similar time period.

The Mk.5 was powered by the 6355lb-ft Orenda 10 and also had the '6-3' wing as standard. Canada built 370 of them and 75 would be supplied to West Germany from 1957 to 1958. These only lasted in service for around four years – most having been disposed of by March 1962.

The final Canadian Sabre was the Mk.6, which had the 7275lb-ft Orenda 14, a lightened airframe and the reintroduction of leading edge slats for improved low-speed handling. This type was retired from front line service in 1963 but would continue to be operated by some RCAF units until December 1968.

South Africa ordered 34 Mk.6s in 1955 and they arrived in 1956. Evidently 17 received Afrikaans markings and went to 1 Squadron and the other 17 were given English markings and went to 2 Squadron. The latter replaced its Sabres with Mirage IIICZs in 1964 but 1 Squadron kept its Mk.6s till 1976 when it replaced them with Mirage F.1AZs.

▼ CANADAIR CL-13 B SABRE MK.6
Canadair CL-13 B Sabre Mk.6, JB-116, *Jagdgeschwader 72*, *Luftwaffe* (72 Fighter Wing, German Air Force), Leck air base, West Germany, 1964.
JG 72 operated the Sabre until 1966 before transitioning to the Fiat G-91.

▼ NORTH AMERICAN F-86F SABRE
North American F-86F Sabre, 3-146, 103rd Fighter Squadron, Imperial Iranian Air Force, Congo, 1964.
The IIAF sent a detachment of four F-86s to assist UN operations in the Congo; Iranian Sabres operated alongside Philippine Sabres and Swedish J-29 Tunnan fighter jets.

Colombia got six Mk.6s in 1956, supplemented by two F-86Fs from Spain in 1963. Four were written off as a result of accidents and the remainder were retired in 1966.

The West German Luftwaffe got 225 Mk.6s and the type became the force's primary day fighter. The first examples were delivered to Jagdgeschwader 71 'Richthofen', commanded by Major Erich Hartmann, the highest scoring ace of all time, on June 6, 1959. The Sabres served for five years till their replacement by F-104Gs and Fiat G-91Rs from 1964 to 1966.

Ninety of Germany's retired Sabres were then transferred to Pakistan in 1966. These would then see action, alongside surplus F-86Fs (see below), against Indian fighters during the 1971 Indo-Pakistan War – accounting for the bulk of the PAF's 141 victories. Among these 'kills' were Hawker Hunters, MiG-21s, Su-7s and a Folland Gnat. A total of 28 Sabres were destroyed during the conflict.

In the aftermath of the war East Pakistan become the independent nation of Bangladesh, and the new Bangladeshi government came into possession of five Pakistani Mk.6s that had been abandoned by the PAF as it retreated. As such, Bangladesh operated the Sabre for two years – having to finally ground its small fleet in 1973 due to lack of spares. Pakistan's remaining Sabres were essentially run into the ground, with their retirement becoming a necessity in 1980 following a series of fatigue-related accidents.

NORTH AMERICAN F-86F SABRE

NORTH AMERICAN F-86F SABRE
North American F-86F Sabre, 709, Blue Impulse aerobatic team, 1st Air Wing, Japanese Air Self Defence Force, Hamamatsu Air Base, Japan, 1977.
The Blue Impulse team flew Sabres from 1960 until 1981; it then transitioned to the Mitsubishi T-2 and later Kawasaki T-4.

US JET FIGHTERS IN FOREIGN SERVICE
North American F-86/CL-13 Sabre

A grand total of 1815 Canadian Sabres were made overall.

EX-USAF F-86FS
When the Korean War ended, the USAF found itself in possession of far more relatively new aircraft than it needed. As such, the F-86F joined the list of American aircraft sent abroad to allies under the Mutual Defense Assistance Program.

Prior to ordering its own Mk.6s, South Africa had operated 22 F-86Fs on loan from the USAF during the Korean War. These were returned when the war ended and joined shipments heading for the Republic of China Air

▼ NORTH AMERICAN F-86F SABRE
North American F-86F Sabre, 832, 7th Tactical Fighter Squadron, *Hukbong Himpapawid ng Pilipinas* (Philippines Air Force), Basa Air Base, Philippines, 1972.
Philippine Air Force F-86 Sabres were used in internal operations against insurgent groups; several demonstration teams were formed within the F-86 fleet.

▼ NORTH AMERICAN F-86F SABRE
North American F-86F Sabre, 180, *Escuadrón Aéreo 111, Grupo Aéreo No. 11, Fuerza Aérea del Perú* (111 Air Squadron, No. 11 Air Group, Peruvian Air Force), Talara Air Base, Peru, 1970.
Peru's Sabres were operated until the late 1970s.

Force in Taiwan. The ROCAF began to receive F-86Fs in December 1954 and by June 1958 had 320, plus seven examples of the reconnaissance variant – the RF-86F. During the Second Taiwan Strait Crisis in 1958, ROCAF F-86Fs engaged in numerous dogfights against Communist MiG-15s and MiG-17s. Equipped with the new AIM-9 Sidewinder, the Fs proved immensely effective against their opponents. During one month, 29 MiGs were confirmed destroyed with eight probables for a loss of two F-84Gs and no Sabres. The Sabre would remain in ROCAF service till 1977.

The Japanese Air Self Defense Force was formed in 1953 and it was decided the following year that the F-86F would become its first standard fighter. Deliveries of surplus Sabres commenced in December 1955 and a total of 135 had been received by early 1957. Beginning in 1956, Mitsubishi built another 300 F-86F-40s under licence. In 1960 a new aerobatic team was formed, Blue Impulse, equipped with Sabres and December 1961 saw Mitsubishi modifying 18 of the American-supplied F-86Fs to RF-86F standard and these were operated till 1979. The last F-86F in Japanese service made its final flight on March 15, 1982.

Pakistan also began to receive F-86Fs in 1954 – eventually being supplied with 120. These saw active service during the 22-day Indo-Pakistani War in 1965 alongside the PAF's Canadair Mk.6 Sabres.

Starting in 1955, Spain began receiving surplus F-86Fs in exchange for letting the USAF use some of its air bases. By the end of 1958 it had received 270, these being locally reconditioned and upgraded. Their replacement with Northrop F-5s began in 1967 and had been completed by the end of 1974.

Four more countries would receive ex-USAF F-86Fs in 1955: Belgium, Peru, Venezuela and the Republic of Korea. Belgium got five but the type was never adopted and the aircraft were presumably returned. Peru got 15 initially and a handful of additional examples some time later. Nine had been wrecked in accidents by December 1963 and the few survivors had been retired by 1980. Venezuela also received a relatively low number of airframes – 30 between 1955 and 1960. Four were used to strafe the presidential palace in Caracas during a failed coup in 1958, six were lost in accidents and the remainder were grounded in 1969. Nine were eventually transferred to Bolivia in 1973.

Incredibly, Bolivia would go on to operate these for another 20 years – finally retiring them in 1993.

Five F-86Fs were handed over to Republic of Korea Air Force pilots on June 20, 1955, and South Korea had received 85 by June 1956. Another 27 plus ten RF-86Fs were delivered during 1958. Many were upgraded to carry AIM-9 Sidewinders before being replaced by F-5s in 1965. The type was eventually retired in 1987.

Norway and the Philippines joined the F-86F club in 1957 with the former receiving 115 from the US and the latter receiving 40 ex-ROCAF examples. Norway grounded its Sabres in the mid-1960s due to wing fatigue but survivors were later handed on to Saudi Arabia and Portugal. The Philippines began taking its F-86Fs off the front line in 1966 but continued to fly them until 1984.

Portugal had already received 50 ex-USAF F-86Fs in 1958 and during August 1961 eight had been deployed to Portuguese Guinea, where they spent three years flying ground-attack and close support missions against rebel forces. The survivors, together with the Norwegian aircraft, would continue to serve until July 1980.

Iran, Iraq and Saudi Arabia would also get Sabres in 1958. Iraq took delivery of just five before the government of King Faisal II was overthrown during a coup; the number received by Iran is unclear and Saudi Arabia got 16. The latter were

US JET FIGHTERS IN FOREIGN SERVICE — North American F-86/CL-13 Sabre

▼ NORTH AMERICAN F-86F SABRE

North American F-86F Sabre, 5313, *Esquadra 51 'Falcões'*, *Força Aérea Portuguesa* (51 Squadron, Portuguese Air Force), aerodrome-base 2, Bissalanca, Guinea-Bissau, 1961.
In 1961, with the perceived threat posed by MiG-17s from neighbouring Guinea-Conakry, the Portuguese Air Force operated a detachment of F-86s in Guinea-Bissau.

▼ NORTH AMERICAN F-86F SABRE

North American F-86F Sabre, 216, Republic of China Air Force, Chiayi Air Base, Taiwan, 1958.
Taiwan operated more than 300 Sabres, being one of the first foreign operators of the aircraft. Taiwanese Sabres were the first aircraft to use the new AIM-9 Sidewinder air-to-air missile in combat.

▼ NORTH AMERICAN F-86 SABRE

North American F-86 Sabre, 603/B, *Ruth II,* 2nd Squadron, South African Air Force, K-55 air base, Osan, Korea, 1953.
The SAAF received F-86s to replace its P-51 Mustangs and operated alongside USAF units during the conflict in Korea.

apparently little used and were retired during the early 1970s.

Ethiopia received 14 F-86Fs in 1960 and Argentina got 28. The latter were used to help defeat an anti-government coup in April 1962 and remained in service long enough to be considered for use against Britain during the Falklands War. They would eventually be retired in 1986. Three Ethiopian F-86Fs were sent to the Congo as part of a UN peacekeeping mission in the early 1960s but didn't stay long, one of them being lost in a crash. The survivors were retired in 1978.

Thailand got 40 F-86Fs in 1962 but they would only serve for four years before being replaced by F-5s. Tunisia then got around a dozen F-86Fs in 1969 and would operate them for nine years – till 1978.

CAC CA-27 SABRE

The Sabres built under licence by the Commonwealth Aircraft Corporation, a subsidiary of British company Hawker, in Australia are generally regarded as the best ever made. The terms of the licence were agreed in October 1951 and Australia placed an order for 72 examples. But a number of changes were specified – the J47 engine would be swapped for an Australian-made Rolls-Royce Avon RA.7, redesignated Avon 20, providing 7500lb of thrust, which meant the nose intake needed to be widened, the fuselage redesigned and the exhaust nozzle enlarged. Armament became a pair of 30mm Aden cannon. Overall, only 40% of the original F-86 structure remained.

The first prototype Australian Sabre, the CA-27 Sabre Mk.30, made its flight debut in August 1953, with the first production machine following in July 1954.

Just 22 Mk.30s were made before the next production version followed – the Mk.31. This had the '6-3' wing modification and late in production an upgraded Avon 20, the Avon 26, was fitted. This created the Mk.32, which then also received the ability to carry a pair of Sidewinders. The last Aussie Sabre was delivered on December 19, 1961, with a total of 112 having been made.

As part of the British Commonwealth Strategic Reserve, Sabres of the RAAF's 3 and 77 Squadrons carried out air strikes against insurgents during the Malayan emergency from February 1959 to July 1960. RAAF Sabres also flew combat air patrols over Thailand in the early stages of what later became the Vietnam war.

The RAAF Sabre fleet was phased out during the late 1960s and early 1970s, with the last example being retired on July 31, 1971. Eighteen Mk.32s were supplied to the Royal Malaysian Air Force between 1969 and 1971 and they in turn were retired in 1978.

Finally, in February 1973 Indonesia bought 18 ex-RAAF Sabres – eventually replacing them with F-5s during the early 1980s.

Nearly 10,000 Sabres were built, excluding the F-86D and its derivatives, making it the most-produced Western jet fighter of all time.

US JET FIGHTERS IN FOREIGN SERVICE

NORTH AMERICAN
F-86D, K AND L SABRE DOG

The F-86D was designed as a radar-equipped interceptor before being simplified for foreign production as the F-86K. These types, along with the upgraded F-86L, continued the Sabre tradition of service with nations around the world.

1955-1993

North American began work on incorporating a radar dish and fire control system into the F-86 design with project NA-164 of March 1949. Initial results were positive and changes to create a production version were crystalised as NA-165 the following month. A mock-up was inspected in June – just three months after the project's inception.

The aircraft's new form was unlikely to win any beauty contests however. The Westinghouse AN/APG-36 18in diameter radar dish was housed in a large new 'nose' which stuck out above a truncated air intake. Other changes included a reshaped fuselage for extra internal fuel capacity, a retractable pod containing 2.75in Mighty Mouse Folding-Fin Aerial Rockets (FFARs) below the cockpit and a clamshell canopy. The fire control system was initially a Hughes E-3, later an E-4.

It was intended that in active service the aircraft would be directed to its target by ground control before the pilot took over and conducted the attack using a radar screen mounted in his main instrument panel.

The first prototype made its flight debut on December 27, 1949, and the type was initially designated F-95A since it shared only 25% of its parts in common with the F-86. However, its designation was changed to F-86D in July 1950 so that Congress would not find itself having to approve funds for a

▼ NORTH AMERICAN F-86D SABRE DOG

North American F-86D Sabre Dog, AL-H/16154, *Eskadrille 726, Flyvevåbnet* (726 Squadron, Royal Danish Air Force), Aalborg Air Base, Denmark, 1960.
This aircraft has the Mighty Mouse rocket pack extended; later these aircraft would have AIM-9 Sidewinder air-to-air missiles.

▼ NORTH AMERICAN/FIAT F-86K SABRE DOG

North American/Fiat F-86K Sabre Dog, 13.QV/ 54860, *Escadron de Chasse 1/13 Artois, Armee de l'Air* (Fighter Squadron 1/13, French Air Force), Air Base 709, Cognac-Châteaubernard, France, 1959.
F-86Ks would be operated by this unit from 1956 until 1962, being later replaced by the Dassault Mirage IIIC.

US JET FIGHTERS IN FOREIGN SERVICE

US JET FIGHTERS IN FOREIGN SERVICE
North American F-86D, K and L Sabre Dog

▼ NORTH AMERICAN/FIAT F-86K SABRE DOG

North American/Fiat F-86K Sabre Dog, JD-337, *Jagdgeschwader 74, Luftwaffe* (Fighter Wing 74, German Air Force), Neuburg Air Base, West Germany, 1965.
The Luftwaffe operated 86 F-86Ks.

▼ NORTH AMERICAN/FIAT F-86K SABRE DOG

North American/Fiat F-86K Sabre Dog, 0019, *Escuadron Caza 35, Fuerza Aérea Venezolana* (Fighter Squadron 35, Venezuelan Air Force), Generalissimo Francisco de Miranda Air Base, Caracas, Venezuela, 1970. Acquired from the Luftwaffe, the F-86K was first all-weather jet fighter of the FAV; it was operated from 1966 until 1969.

040 US JET FIGHTERS IN FOREIGN SERVICE

▼ NORTH AMERICAN/FIAT F-86K SABRE DOG

North American/Fiat F-86K Sabre Dog, 51-31/ 54891, *22º Gruppo, 51ª Aerobrigata, Aeronautica Militare* (22nd Group, 51st Air Brigade, Italian Air Force), Istrana Air Base, Italy, 1962.
Italy used both domestically built Fiat F-86Ks and imported North American examples; this aircraft carries AIM-9B Sidewinder air-to-air missiles.

whole 'new' aircraft rather than a variant of an existing one.

The rocket pod was first test-fired in February 1951 and the first production F-86D flew on June 8 of that year. Its service ceiling was 55,400ft and it therefore required a system where warm air from the engine was piped to the wings, fin and intake leading edges to prevent them from icing up. Air Defense Command was so impressed with the F-86D's speed and altitude performance that it opted to equip two thirds of its wings with them.

The USAF received its first F-86D for testing on March 12, 1952, and the first front-line fighter squadron to receive it did so in February 1953. Less than a month earlier, it had been decided that Italy would build F-86Ds under licence – but with a simpler fire control system using less sensitive cutting-edge tech and cannon instead of FFARs. It would be a two-seater too. North American successfully argued against the latter stipulation and project NA-205 was created as a single-seater with a basic MG-4 fire control system for four 20mm M24A1 cannon. It kept the APG-37 radar, which allowed the new aircraft to fly essentially the same mission profile as the American original.

Two existing F-86Ds were converted as prototypes for the NA-205 configuration under the designation YF-86K. Italian firm Fiat signed the licence agreement with North American on May 16, 1953, which would see an initial batch of 50 F-86Ks assembled in Italy using US-made components. Since the aircraft were urgently required, not just by Italy but also France, the new West German Luftwaffe, Norway and the Netherlands, it was agreed on December 18, 1953, that North American itself would also build 120 F-86Ks and ship them over to the latter two nations.

The first YF-86K, with a nose 6in longer than that of the F-86 to provide more space for the cannon, flew on July 15, 1954. It had vents on the gun bay doors to prevent gun gas build-up but in most other respects was visually similar to the F-86D. Following initial testing both YF-86Ks were shipped to Italy. The first US-made F-86K flew on March 8, 1955, followed by the first Italian-assembled example on May 23, 1955. The 120 American F-86Ks were made from April to December 1955 – 60 being earmarked for Norway and 59 for the Netherlands; North American retained a single aircraft for itself. Deliveries to Norway commenced in September 1955 and continued into October 1956. One was lost during acceptance trials in the US but was replaced in January 1960. Four more were destroyed during a hangar fire on March 10, 1956 and these were replaced with Italian F-86Ks. The Norwegian fleet was withdrawn and scrapped between 1966 and 1967.

Holland's 59 F-86Ks were delivered from the US between October 1955 and April 1956. An additional six Italian F-86Ks were added in 1957. Eight were

US JET FIGHTERS IN FOREIGN SERVICE
North American F-86D, K and L Sabre Dog

▼ NORTH AMERICAN/FIAT F-86K SABRE DOG
North American/Fiat F-86K Sabre Dog, 1101, *Fuerza Aérea Hondureña* (Honduran Air Force), Coronel Armando Escalón Espinal Air Base, Honduras, 1976.
Honduras acquired five examples from Venezuela.

▼ NORTH AMERICAN F-86L SABRE DOG
North American F-86L Sabre Dog, 1223/30853, 12th Squadron, Wing 1, Royal Thai Air Force, Don Muang air base, Thailand, 1966.
Thailand was the only foreign operator of the F-86L variant, featuring improvements in aerodynamics, engine and avionics.

returned to Italy in 1963 and a total of 16 were written off by the time the F-86K was replaced in Dutch service by F-104G Starfighters during the mid-1960s. The remainder were mostly scrapped.

After the initial 50, North American then sent another 171 F-86Ks to Fiat in kit form for a total of 221 delivered between August 1954 and December 1955. Deliveries of 63 fully assembled aircraft to the Italian air force commenced on November 2, 1955. The last one was delivered in October 1957 and some would serve until July 1973, despite the official phaseout and replacement with F-104Gs having commenced in 1964.

Meanwhile, starting in May 1956, a total of 981 American F-86Ds had been converted into F-86Ls with the installation of Semi Automatic Ground Environment or SAGE equipment – allowing ground control to communicate data directly to the FCS rather than having to relay it vocally to the pilot. The aircraft also received radar upgrades, extended wingtips and wing leading edges, a revised cockpit layout and an uprated engine.

Fiat commenced F-86K deliveries to France in September 1956 and continued into mid-1957, with a total of 62 sent. These aircraft would be replaced with Mirage IIICs by August 1962 and 22 were then returned to Italy, most of the remainder being destroyed. The Luftwaffe got 86 Italian-assembled F-86Ks between 1957 and 1958 and all were made Sidewinder capable and had Martin-Baker ejection seats fitted in 1962. They would eventually be phased out and replaced with F-104Gs in 1966.

But rather than being scrapped, 74 ex-German F-86Ks were then sold to Venezuela along with others which served as a source of spares. In fact, all but 27 would eventually be cannibalised for parts and the survivors were grounded in July 1969 due to worn out hydraulic hoses.

Five of those aircraft were then sold to Honduras and may have seen action not long after the 1969 Football War against El Salvador. They were withdrawn from service in 1980.

Finally, Fiat overhauled 40 F-86Ks between 1963 and 1964 before sending them to Turkey where they would remain in service till 1969.

F-86D AND F-86L

The E-4 fire control system was finally deemed safe for export in 1958 and as such retired USAF F-86Ds could now serve with the air forces of US allies around the world. Denmark got 59 that year with a further three serving as a source of spare parts. They would be withdrawn from front line units in March 1966 but some would continue to fly as decoys. Japan received the first of 122 F-86Ds in 1958 and operated the type for a decade. The Philippines too would get F-86Ds in 1958 – though only 20. These had all been retired by 1970. Taiwan received 25 F-86Ds – enough for one all-weather interceptor squadron – and was still operating them by 1965. Further detail on these aircraft is lacking.

Three more nations would get F-86Ds in 1961; Greece received 35 and flew them in a single unit, 343 Squadron, until 1967 – with a handful continuing to operate as decoys until, incredibly, 1993. Turkey got 50 at the same time and Yugoslavia got no fewer than 130, where they served alongside MiG-21s. They were finally retired in 1980.

The only foreign operator of ex-USAF F-86Ls was Thailand. The Royal Thai Air Force received 17 examples minus their SAGE gear, which was stripped out before delivery, during 1961-1962. These would continue to serve into the early-to-mid-1970s.

▼ NORTH AMERICAN F-86D SABRE DOG

North American F-86D Sabre Dog, 088, *117. lovački avijacijski puk, Ratno vazduhoplovstvo i protivvazdušna odbrana* (117 Fighter Aviation Regiment, Yugoslav Air Force and Air Defence), Pleso Air Base, Yugoslavia, 1967.
Sabre Dogs provided all-weather fighter capability to Yugoslavia, operating alongside MiG-21s.

US JET FIGHTERS IN FOREIGN SERVICE

NORTH AMERICAN F-100 SUPER SABRE

The tough and capable Super Sabre was operated by four foreign nations and would see combat with three of them.

North American started work on a supersonic Sabre in February 1949 by focusing on three key modifications – increasing wing sweepback from 35° to 45°, an area-ruled fuselage and a more powerful turbojet.

These design elements featured in a series of proposals submitted by the company to the USAF, each being rejected and the company then trying again with further alterations. By the third proposal, the NA-180 Sabre 45, very little of the original F-86 remained. The aircraft was bigger, heavier and more powerful, incorporating the Pratt & Whitney J57-P-7 Turbo Wasp afterburning turbojet.

The USAF decided to buy the new type in October 1951 and placed an order for two prototypes plus 110 full production models the following month. The mock-up inspection on November 9 resulted in requests for more than 100 changes – removing any last vestiges of the F-86 by extending

1958-1982

▼ NORTH AMERICAN F-100D SUPER SABRE

North American F-100D Super Sabre, 11-YF/42130, *Escadron de Chasse 4/11 'Jura', Armee de l'Air* (4/11 Fighter Squadron, French Air Force), Air Base 188, Djibouti, 1978.
The last French squadron to operate the F-100s, EC 4/11 was also the only permanently overseas based fighter squadron of the French Air Force.

the cockpit canopy, lowering the tailplanes and reshaping the fuselage. A month later the type received the new F-100 designation.

More than six months later additional changes were made – the self-sealing fuel tanks were to be swapped for lighter non-self-sealing tanks, the aircraft's nose was lengthened by 9in, the tailplanes and fin were made slightly shorter with increased chord and fixings for external weapons racks being added. Development of wing fuel tanks for the type began in October 1952 and the first YF-100A prototype made its flight debut on May 25, 1953.

Above 30,000ft the YF-100A could go supersonic with ease and could fly close to Mach 1 even at low level. Less than two months later, on July 6, 1953, the aircraft hit Mach 1.44 while diving from an altitude of 51,000ft. That same month, the USAF followed up on its request for wing fuel tanks with a request that these same new wings should be able to carry a bomb payload.

The second YF-100A flew on October 14, 1953, and the first production example flew a week later. Colonel Frank Kendall 'Pete' Everest set a new world air speed record of 755.149mph in the aircraft on the same day. Early on, pilots found that poor visibility from the cockpit made take-offs and landings tricky and the sharply swept wings meant landing speed was necessarily high. Similarly, the type was a handful when flying at low speed and longitudinal stability was lacking at high speed. Some rudder

▼ NORTH AMERICAN F-100F SUPER SABRE
North American F-100F Super Sabre, GT-927, *730 Eskadrille, Flyvevåbnet*, (730 Squadron, Royal Danish Air Force), Skrydstrup Air Base, Jutland, Denmark, 1981.
Danish Super Sabres would be operated until the early 1980s.

US JET FIGHTERS IN FOREIGN SERVICE North American F-100 Super Sabre

flutter was present too, but this was corrected using hydraulic dampers.

Deliveries to the USAF commenced at the end of November 1953 and a month later the USAF decided that the last 70 F-100As should be modified to become NA-214 fighter-bombers with the new fuel and bomb load-carrying wing – later to be redesignated F-100Cs. By May 27, 1954, the number of F-100Cs on order was 564.

On September 27, 1954, an order was made to complete many of the pre-ordered F-100Cs as F-100Ds instead. The F-100D had another new wing – this time with a greater chord at the wingroot due to less swept inner flaps which increased surface area and lowered landing speed. The underwing pylons could be jettisoned using explosive bolts rather than relying on gravity and a new centreline hardpoint was introduced. The F-100D also featured built-in electronic countermeasures equipment, the AN/AJB-1 low-altitude bombing system and an AN/APS-54 tail warning radar.

Following a series of mid-air disintegrations, all F-100s were fitted with a lengthened fin with 27% more area. North American received a contract to build a two-seat trainer F-100, the TF-100C, in December 1955, and a full production two-seater, the F-100F, had entered production by the end of 1957.

The F-100's first export customer was France. Deliveries started in May 1958 and it would eventually receive 85 F-100Ds plus 15 F-100Fs. Since France had been fighting insurgents in its colony of Algeria since November 1954, these new aircraft were quickly pressed into service. EC 1/3 Navarre flew bombing missions from its base in Rheims against rebel forces until 1962 when the colony finally won its independence. The French Super Sabres served for two decades before their withdrawal in 1978.

Turkey began to receive F-100Cs, Ds and Fs during the late 1950s and these

▼ NORTH AMERICAN F-100D SUPER SABRE

North American F-100D Super Sabre, FW-242, *171 Filo, Türk Hava Kuvvetleri,* (171 Squadron, Turkish Air Force), 7th main air jet base, Erhaç, Turkey, 1976.
Turkish F-100s saw action during the 1974 Cyprus conflict.

too would conduct combat operations – against Greek forces during conflicts over Cyprus in 1964 and again in 1974. On August 8, 1964, an F-100D flown by Cengiz Topel was shot down by 40mm anti-aircraft fire during the Battle of Kokkina. Topel ejected over land and was captured before being tortured to death. During the battles of 1974, three Turkish F-100Cs and three F-100Ds were shot down. The surviving aircraft were retired in 1982.

Taiwan received a single F-100F in October 1958, then 15 F-100As in 1959 and another 65 F-100As in 1960. Four unarmed RF-100A reconnaissance aircraft followed in 1961 and then 38 more F-100As after that. Most of these had been retrofitted with the larger tail fin, AN/APS-54 tail-warning radar and Sidewinder capability.

The RF-100As were reportedly used for deep penetration overflights of the Chinese mainland in the 1960s and it is believed that at least one failed to return. Full details of these operations remain undisclosed.

The fourth foreign F-100 operator was Denmark, which received 48 F-100Ds and ten F-100Fs, starting in July 1959. While these never saw action, they did suffer a very high rate of attrition – a third of the fleet being lost in accidents. The remainder were retired during the early 1980s and replaced with F-16As.

▼ NORTH AMERICAN F-100A SUPER SABRE

North American F-100A Super Sabre, 0228, 2nd Tactical Fighter Wing, Republic of China Air Force, Hsinchu Air Base, Taiwan, 1959. Taiwan was the only foreign operator of the F-100A variant.

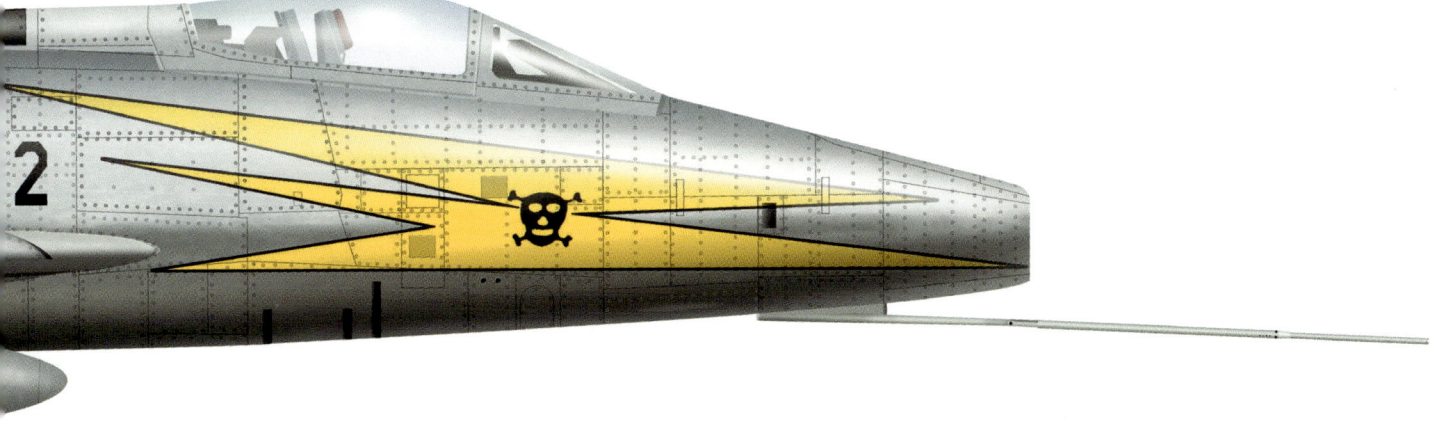

US JET FIGHTERS IN FOREIGN SERVICE

MCDONNELL F-101 VOODOO

Enormous and difficult to fly, the F-101 was nevertheless a fast and powerful interceptor as well as being a capable reconnaissance platform. Just two countries would operate it besides the US.

The F-101 Voodoo was a much-improved development of the McDonnell F-88, originally designed to meet an early 1946 requirement for a long-range bomber escort. Work on the F-88 had been protracted but it was finally ordered into production in October 1951, being redesignated F-101 at the end of the following month.

As development continued, the original design was lengthened to include a pair of Pratt & Whitney J57-P-13 engines and additional fuel. Its intakes were enlarged, the tail planes were repositioned higher up the fin and the wings were increased in chord. Armament was four 20mm M39 cannon with a K-19 gunsight.

The mock-up was inspected in July 1952 and 39 F-101As were ordered on May 28, 1953. The USAF had asked McDonnell to develop a reconnaissance version, the RF-101A, in January 1953. This would have a longer nose housing four low-altitude cameras plus two high-altitude cameras positioned behind the cockpit.

Next, in June 1954, the USAF ordered a two-seat all-weather interceptor version as the F-101B. This would have a Hughes

1959-1987

▼ **MCDONNELL RF-101A VOODOO**
McDonnell RF-101A Voodoo, 5656/41518, Republic of China Air Force, Taoyuan air base, Taiwan, 1960.
Taiwanese Voodoos were used in reconnaissance missions over mainland China and some were shot down.

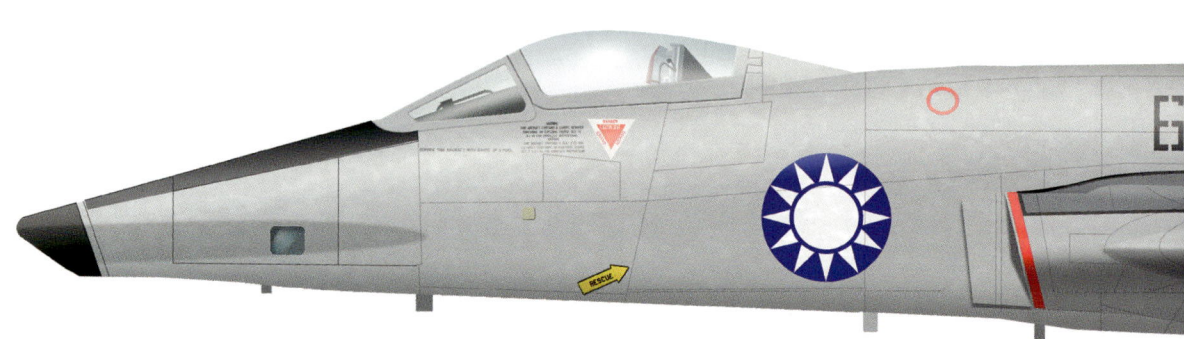

MG-13 fire control system and would be armed with two Falcon air-to-air missiles and two Genie unguided nuclear rockets.

The F-101A's first flight was on September 29, 1954, the YRF-101A first flew on June 30, 1955, and the first F-101B flew on March 27, 1957. Produced alongside the B was the F-101F – an externally identical two-seat operational and conversion trainer. The F had dual controls but was also fully combat capable.

Taiwan began to receive its first Voodoos, ex-USAF RF-101As, in November 1959 under Operation Boom Town. It would eventually receive a total of eight and the Republic of China Air Force flew them on spy missions over the Chinese mainland. During these hazardous operations, two would be shot down by MiGs and another by AAA fire. The remainder would be stood down in the late 70s.

Even as Taiwan was adopting the Voodoo, Canada was cancelling its indigenous Mach 1.9 all-weather interceptor – the CF-105 Avro Arrow. It had intended to rely on IM-99 Bomarc missiles for defence instead but it soon became clear that this had been an extremely bad move. Canada therefore decided it needed Voodoos to fill what was now a large gap in Royal Canadian Air Force capabilities.

Fifty-six former USAF F-101Bs and ten F-101Fs were delivered under Operation Queen's Row from July 1961 to May 1962 and Canada redesignated them as CF-101Bs and CF-101Fs. They carried the same missile load as their American counterparts. Nearly a decade later, in 1970-1971, there were only 46 CF-101s remaining. These were then traded back to the USAF in exchange for another 56 F-101Bs and another 10 CF-101Fs. Refurbished and modernised, this second batch of aircraft were also equipped with infrared sensors and improved fire control systems. One F-101B, serial 101067, was reequipped as an electronic countermeasures aircraft and continued to serve after all others, except one, had been retired in 1985. The other F-101B, 101006, would serve as a trainer for 101067. Both were finally withdrawn from service in 1987.

▼ MCDONNELL F-101B VOODOO

McDonnell F-101B Voodoo, 101057, 409 All-Weather Fighter Squadron, Royal Canadian Air Force/Aviation Royale Canadienne, Canadian Forces Base Comox, Canada, 1984.
To commemorate the retirement of the Voodoo from Canadian service, each squadron painted one of its aircraft in a special scheme; *Hawk One Canada* was the contribution of 409 Squadron.

US JET FIGHTERS IN FOREIGN SERVICE **049**

US JET FIGHTERS IN FOREIGN SERVICE

CONVAIR F-102 DELTA DAGGER

Designed and built to intercept Soviet bombers heading for the US, the Delta Dagger was not an obvious candidate for export – yet it would nevertheless be operated by bitter opponents Turkey and Greece.

The F-102 was the product of a new concept in fighter technology – the 'weapons system'. The goal was to simultaneously develop the aircraft and its fire control system, rather than each being developed without reference to the other.

Starting in January 1950, companies were invited to bid for contracts to design a new interceptor and its electronics. Eventually a futuristic pure delta from Convair was chosen as the airframe while Hughes beat North American to the electronics contract. The aircraft received the designation XF-102. The first prototype, designated YF-102, flew on October 23, 1953.

Many development difficulties followed, with the wasp-waisted, J57-P-41-powered production model F-102A eventually flying on June 24, 1955. Flutter problems then needed to be resolved before the type finally entered

▼ CONVAIR F-102 DELTA DAGGER

Convair F-102 Delta Dagger, 1974, 0-53413, *142 Filo, Türk Hava Kuvvetleri,* (142 Squadron, Turkish Air Force), 4th Main Jet Air Base, Mürted, Turkey, 1974.
Turkish F-102s were used during the 1974 Cyprus conflict, reportedly engaging Greek F-5s.

1968-1979

operational service with the USAF in April 1956. A two-seat trainer variant was also produced as the TF-102A.

F-102As and TF-102As were provided to both Turkey and Greece under the Peace Violet programme, commencing in late 1966. Six Turkish instructors began training on F-102s at Perrin AFB, Texas, in November 1967 and presumably a similar course was also provided for their Greek counterparts. Prior to delivery, aircraft destined for service abroad were initially sent to subcontractor CASA at Seville in Spain to undergo basic refurbishment. Deliveries of 49 F-102As and nine TF-102As to Turkey began in 1968 and had been completed by 1971. The Hellenic Air Force received the first of 19 F-102As and five TF-102As in June 1969.

The F-102A had a fairly unremarkable career with both nations until 1974 when Turkey invaded Cyprus on July 20, following a coup ordered by the military junta in Greece on July 15. While there remains no evidence of F-102A fighting F-102A, it is claimed that Greek F-5As shot down two Turkish F-102As – one with a Sidewinder and the other with cannon fire. Turkey claimed that one of its F-102As had shot down a Greek F-5A.

Shortly after the conflict, the US placed an arms embargo on Turkey which made its fleet of US-made equipment increasingly difficult to maintain. The F-102A fleet, which had already been subject to a high rate of attrition, suffered particularly badly and the type was withdrawn from service in 1979. Greece had retired its F-102s even earlier, in 1978, and replaced them with Mirage F1CGs.

▼ CONVAIR TF-102 DELTA DAGGER

Convair TF-102 Delta Dagger, 0-62327, *342 MPK, Polemikí Aeroporía* (342 All-Weather Intercept Squadron, Hellenic Air Force), Tanagra Air Base, Greece, 1976.
The two-seat variant had also a weapons bay and retained combat capability.

US JET FIGHTERS IN FOREIGN SERVICE

LOCKHEED F-104 STARFIGHTER

It was unwanted by the USAF but Lockheed's Starfighter offered staggering performance and went on to massive worldwide success with the infamous 'sale of the century'.

▼ LOCKHEED F-104G STARFIGHTER

Lockheed F-104G Starfighter, FX51, 350th Squadron, 1st Wing, Slivers Demonstration Duo, *Belgische Luchtmacht – Force Aérienne Belge (*Belgian Air Force), Beauvechain Air Base, Belgium, 1971.
From 1969 to 1975, two pilots from the 350th squadron, formed a demonstration duo using the F-104, flying both at home and abroad. This aircraft displays the original team's emblem on its intake, soon replaced by the team's name.

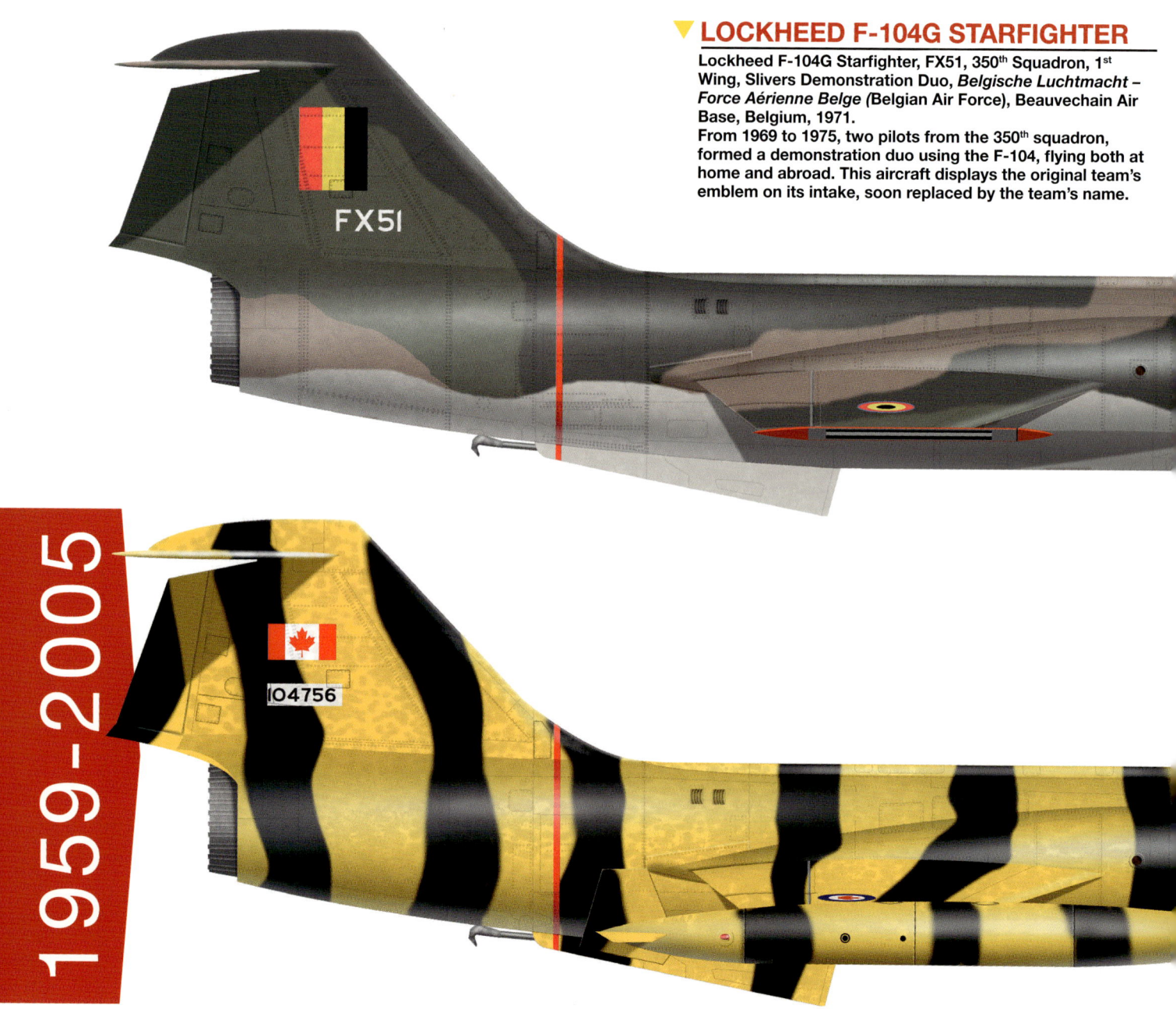

1959–2005

The design that would become the F-104 was submitted to the USAF in early 1953 for the WA-303A weapon system requirement. Originally designated L-246 or Model 83, it featured a long slender fuselage tapering to a pointed nose and extremely short, thin, unswept wings. Famously, the F-104's wing leading edge had a radius of just 0.016in and had to have a protective cover to prevent ground crew from being injured by it.

The USAF ordered a pair of XF-104 prototypes in March 1953 and the first one took to the air briefly on February 28, 1954, before making a full flight on March 4. Powered by a Curtiss-Wright XJ65-W-6 it could manage Mach 1.79. Seventeen pre-production YF-104As came next with deliveries commencing on February 17, 1956. These were fitted with the more powerful General Electric J79-GE-3A which, combined with a taller fin, new spine and variable intakes allowed the type to exceed Mach 2 in level flight for the first time on April 27, 1955.

By October 1956 a total of 153 F-104As, 18 RF-104A reconnaissance variants, 56 F-104C fighter-bombers and 26 F-104B two-seat trainers had been ordered. These would feature new ventral fins and in-flight refuelling gear. Unfortunately, the early J79 engine was unreliable and the aircraft was involved in a series of accidents, with 21 pilots losing their lives testing it. The F-104's unusual Lockheed-designed ejection seat, which fired the pilot down through the fuselage, only made matters worse.

After just a few months in service the F-104A was grounded due to concerns about its mechanical issues and poor handling in flight. In addition, the afterburner could not be regulated so pilots has a choice of flying at Mach 1 without afterburner or Mach 2.2 with afterburner and nothing in between. The RF-104A order was cancelled in 1957 but the F-104C order went ahead. The 'C' had a conventional ejection seat as standard and was designed to drop tactical nuclear weapons. A trainer version of the 'C', the F-104D, was ordered in 1957.

Deliveries of the F-104C began in October 1958. However, lack of faith in the early F-104 design was so great that the F-104As were withdrawn from USAF service in 1959 and transferred to the Air National Guard. Only 296 were delivered for US service in total.

But the F-104's story was only just beginning.

EUROPEAN PRODUCTION
Europe's NATO air forces, besides Britain and France, found that by the mid-1950s they needed a new supersonic multirole fighter able to deliver the US-made B-43 tactical nuclear weapon. All those F-84s and F-86s would soon need to be replaced with something more modern. It became apparent to the Western aircraft manufacturers that this need amounted to a potential market for more than 2000 new fighter-bombers.

Ten contenders were put forward by aircraft manufacturers in England, France, Sweden, and the USA: the English Electric Lightning, Saunders-Roe SR.177, Dassault Mirage III, SAAB J-35 Draken, Convair F-102 Delta Dagger, Convair F-106 Delta Dart, Republic F-105 Thunderchief, Vought F8U Crusader, Grumman F11F-1F Tiger, and Lockheed F-104 Starfighter. It was widely believed that once the West German Luftwaffe had made its choice, the other customer nations would follow suit. Whoever won the contract would enjoy the 'sale of the century'.

Lockheed's proposal, the F-104G (G for Germany) was based on the F-104C but significantly upgraded. It would have all-weather air-to-air and air-to-ground combat capability thanks to an Autonetics F15A NASARR (North American Search and Ranging Radar) fire control system, director gunsight with infrared optics, Litton LN-3 inertial navigator, upwards-firing ejection seat, larger undercarriage with beefier brakes, strengthened fuselage, wing and tail.

▼ CANADAIR CF-104 STARFIGHTER
Canadair CF-104 Starfighter, 104756, No.439 Squadron, Royal Canadian Air Force/Aviation Royale Canadienne, Baden–Soellingen Air Base, West Germany, 1976.
Properly painted for the occasion, this Germany-based RCAF Starfighter participated in the 1976 NATO Tiger Meet.

US JET FIGHTERS IN FOREIGN SERVICE — Lockheed F-104 Starfighter

The tail was the larger unit fitted to the F-104B/D, including a fully-powered rudder. Internal fuel capacity was increased and five hardpoints could carry 4000lb of stores. The engine was General Electric's J79-GE-11A, providing 10,000lb thrust dry or 15,600lb with afterburner.

A key factor in Germany's decision to choose the F-104G was the opportunity to build it under licence in Germany – generating jobs and expanding the German national aircraft industry. A contract for 66 F-104Gs was awarded to Lockheed on February 6, 1959, rising to 96 shortly thereafter.

A consortium of German aircraft manufacturers acquired the licence to manufacture the Starfighter on March 18, 1959. Canada then also selected the Starfighter on July 2, 1959, with Canadair lined up to build it locally. The Netherlands completed a licensing agreement on April 20, 1960. Belgium signed a similar

▼ LOCKHEED F-104G STARFIGHTER

Lockheed F-104G Starfighter, 26+69, *Marinefliegergeschwader 2, Marineflieger, (Aero-naval Wing 2,* German Navy Air Arm), Ramstein Air Base, Rhineland-Palatinate, West Germany, 1984.
German Navy's F-104 could be equipped with the AS.34 Kormoran anti-ship missile.

▼ AERITALIA F-104S ASA STARFIGHTER

Aeritalia F-104S ASA Starfighter, MM6945, *311°Gruppo, Reparto Sperimentale Volo, Aeronautica Militare* (311 group, Experimental Flight Department, Italian Air Force), Mario de Bernardi military airport, Pratica di Mare, Italy 1984.
The prototype for the F-104S ASA is seen here carrying both AIM-9 Sidewinder and Aspide air-to-air missiles.

agreement on June 20 and Italy signed up on March 2, 1962. The J79 would be built under licence by MAN-Turbo in Germany, Fabrique Nationale in Belgium and Fiat in Italy. Meanwhile, Japan had decided to adopt the F-104 as its standard air superiority fighter in November 1960.

The European producers were joined together in four geographical groups – Dornier, Heinkel, Messerschmitt, Siebel and BMW in the South Group (contracted for 210 aircraft initially); Fokker, Dordrech, Aviolanda, Focke-Wulf, Hamburger Flugzeugbau and Weserflugzeugbau in the North Group (350 aircraft); SABCA, Fairey S.A. and Fabrique Nationale in the West Group (188 aircraft). The fourth group was based entirely in Italy – Fiat, Aerfer-Macchi, Piaggio, SACA and SIAI-Marchetti (199 aircraft). In addition, both Canadair and Lockheed would supply additional components.

Lockheed itself built the first 66 F-104Gs for the Luftwaffe and another 84 for various USAF Mutual Aid contracts. The first of these American F-104Gs flew on June 7, 1960, followed by the first from South Group on October 5, 1960. West Group's first Starfighter made its debut flight on August 3, 1961, and North Group's followed on November 11, 1961. The first Italian F-104G flew on June 9, 1962. Italy would also build a number of RF-104Gs. Beginning in 1967, all remaining European F-104Gs were retrofitted with Martin-Baker Mk GQ7(F) 'zero-zero' ejection seats.

West Germany would operate 35% of all Starfighters built – flying F-104s from all four European groups plus examples from the US; a total of 915 aircraft including new production runs to replace the many aircraft lost in accidents. The German Navy's Marineflieger were allocated 151 F-104Gs and the remainder went to the Luftwaffe. Pilot conversion training began in July 1960 and the first Luftwaffe F-104G unit became fully operational in 1963. The Luftwaffe began phasing the F-104s out in 1971, gradually replacing them with F-4s. The German Navy began converting to the Panavia Tornado in 1982. The last Starfighter in German front-line service was retired in October 1987 and the final German Starfighter flight was on May 22, 1991.

The Netherlands received a total of 138 Starfighters – 95 F-104G/RF-104G from the North Group, 25 F-104Gs from the Italian group and 18 TF-104Gs from Lockheed in the US. They entered service in 1962 and had all been replaced with F-16s by the end of 1984.

The Belgian Air Force got 100 F-104Gs from the West Group plus 12 TF-104Gs from Lockheed, starting in February 1963, and had replaced them all with F-16s by September 1983.

Latecomer Spain received 18 F-104Gs and three TL-104Gs built by Lockheed in 1965. They were phased out in May 1972 and replaced by F-4C Phantoms.

▼ LOCKHEED F-104G STARFIGHTER
Lockheed F-104G Starfighter, 7151, *336 Mira, Polemikí Aeroporía* (336 Squadron, Hellenic Air Force), Araxos Air Base, Greece, 1993.
Depicting the elements of the 336 Squadron's emblem, this aircraft nicknamed *Olympus*, was later displayed at the Hellenic Air Force Museum.

CANADIAN PRODUCTION
The Canadian F-104 licensing agreement was signed on September 17, 1959, with Canadair to build 200 CF-104s for the Royal Canadian Air Force. The CF-104 differed from the F-104G in being optimised for nuclear strike rather than air-to-air and air-to-ground operations.

As such, it was fitted with the R-24A NASARR system, which was air-to-ground only, instead of the F15A NASARR. It had an even tougher undercarriage, with larger mainwheel tyres, the removeable refuelling probe of the F-104Cs and Ds, the ability to carry a ventral reconnaissance pod containing four cameras, and a fuel cell in place of the M61A1 cannon – at least initially. Orenda Engines was licensed to build the J79 engine.

The first Canadair-made CF-104 was airlifted to Palmdale, California, for its first flight on May 26, 1961. The first two CF-104s to fly in Canada did so at Montreal on August 14 that year. The

US JET FIGHTERS IN FOREIGN SERVICE — Lockheed F-104 Starfighter

▼ LOCKHEED F-104J STARFIGHTER

Lockheed F-104J Starfighter, 48-8641, 203 Squadron, Japanese Air Self Defense Force, Chitose air base, Hokkaido, Japan, 1979.
Japanese Starfighters were painted with areas of bright colour when participating in air combat exercises.

▼ LOCKHEED F-104A STARFIGHTER

Lockheed F-104A Starfighter, 910/H, 9 Squadron, Royal Jordanian Air Force, Prince Hassan Air Base, Amman, Jordan, 1972.
Jordan operated around 30 examples of the F-104A and B variants; these would be replaced by Northrop F-5s and Dassault Mirage F-1s.

▼ LOCKHEED F-104G STARFIGHTER
Lockheed F-104G Starfighter, D-8331, 312 Squadron, *Koninklijke Luchtmacht* **(Royal Netherlands Air Force), Twente Air Base, the Netherlands, 1980.**
Adorned with a rather happy shark mouth, this RNLAF's F-104 was the demonstration aircraft during the 1979 Air Force Day.

▼ LOCKHEED F-104G STARFIGHTER
Lockheed F-104G Starfighter, 26+44, *Jagdbombergeschwader 33, Luftwaffe,* **(Fighter-Bomber Wing 33, German Air Force), Büchel Air Base, Germany, 1986.**
Germany received more than 900 Starfighters, equipping both *Luftwaffe* and *Marineflieger* units; this aircraft displays the Norm 83 colour scheme.

200th and last CF-104 was finished on September 4, 1963, and delivered to the RCAF on January 10, 1964. Canadair then switched to producing a total of 140 F-104Gs for Norway, Denmark, Greece, Turkey and Spain. Canada also received 38 two-seat trainers from Lockheed, designated CF-104D.

The CF-104s in RCAF service were converted from nuclear to conventional ground attack from 1970 to 1972, being retrofitted with a 20mm Vulcan cannon, twin bomb ejector racks and multi-tube rocket launchers. Starting in 1983, the RCAF began replacing its Starfighters with CF-18 Hornets. The last were phased out in 1986. Around 110 CF-104/CF-104Ds were wrecked in accidents out of 238 – around 46%. But Canadian CF-104s had particularly hard lives, with the average airframe having 6000 hours on it at retirement, compared to 2000 for a Luftwaffe F-104G.

Starting in 1963, Norway received 16 F-104Gs and two TF-104Gs made by Lockheed plus a trio of F-104Gs from Canadair. The US-made F-104Gs were converted to RF-104G configuration before being converted back to fighters. It later got two ex-Luftwaffe F-104Gs and then in 1973 received 18 CF-104s plus four CF-104Ds, the latter 22 aircraft being equipped to fire Martin Bullpup missiles in the antishipping role. All had been phased out by early 1983.

JAPANESE PRODUCTION
An industrial group led by Mitsubishi Heavy Industries built the Japanese F-104 variant – the F-104J. The first few were assembled in Japan from American-made parts but soon Japan was making all parts itself, including the J79-IHI-11A engine which was built under licence by Ishikawajima-Harima. They were fitted with NASARR F-15J-31 fire control systems, optimised for air-to-air combat, and were armed with a 20mm M61A1 cannon plus four Sidewinders.

The first F-104J flew on June 30, 1961, and a total of 178 were delivered from March 1965 to 1967. As with Canada, Lockheed supplied the two-seat trainer – providing 20 kits to be assembled in Japan between July 1962 and January 1964. The F-104J entered front line service with the Japanese Air Self Defense Force in October 1966 and their retirement commenced in December 1981, their replacements being Mitsubishi-made F-15J/F-15DJ Eagles. Around 34 F-104Js and two F-104DJs were lost in accidents – about 15% of the fleet.

PAKISTAN AND TAIWAN
Pakistan received ten ex-USAF F-104As and two F-104Bs in 1961. Four years later during the war with India, these aircraft would fly 246 sorties and claim four kills for the loss of two F-104As. Only seven Pakistani F-104s were still flying by the time war broke out again on December 7, 1971, and this time there were no kills but two losses to Indian MiG-21s with another loss to ground

US JET FIGHTERS IN FOREIGN SERVICE — Lockheed F-104 Starfighter

fire. The survivors would be replaced by Mirage 5Pas in 1975.

The Republic of China Air Force received 24 ex-USAF F-104As and five F-104Bs in 1960-61, then 46 Lockheed-made F-104Gs, eight TF-104Gs and 21 Canadair RF-104Gs in 1964-69. Six F-104Ds followed from US ANG units in 1975. By 1987, the RoCAF had received 22 F-104Js and five F-104DJs from Japan plus 15 F-104Gs and three TF-104Gs from Denmark. In 1967, two RoCAF Starfighters shot down a pair of Chinese MiG-19s – the first Starfighter air-to-air victory. The last Taiwanese F-104s were retired in 1998.

FURTHER ITALIAN PRODUCTION

The Italian Air Force decided it needed a new all-weather interceptor in 1965 and an upgraded F-104 was chosen as the winning design. The Lockheed CL-980 would become the ultimate Starfighter development under the designation F-104S – the 'S' standing for Sparrow, since it could carry AIM-7 Sparrow radar-guided missiles. It was powered by the J79-GE-19 with 11,870lb of dry thrust or 17,900lb with afterburner. This required auxiliary inlet doors on the sides of the intakes to provide extra air during takeoff.

The S also had NASARR R-21G/H fire control for its Sparrows – the earlier F-104s having only been able to carry infrared homing missiles – which incorporated contour/ground mapping plus terrain avoidance. It had nine hardpoints – two on the wingtips, four under the wings, two under the forward fuselage and one on the fuselage centreline for a total payload of up to 7500lb – but the M61A1 cannon was deleted. The ventral fin was enlarged and supplemented by two additional fins, one on either side.

Lockheed modified two Fiat F-104Gs to create F-104S prototypes, the first flying in December 1966, and the first Italian-made example flew on December 30, 1968. The manufacturing group, led by Fiat and including Alfa Romeo, Macchi, GE International, Selenia and FIAR, made 206 F-104Ss for the Italian Air Force and 40 for Turkey. The last deliveries were in March 1979 – marking the end of Starfighter production overall. During the 1980s, 147 Italian F-104S airframes were upgraded to F-104S-ASA standard with a new FIAR R21G/M1 Setter radar. The last Italian F-104 was retired in 2005.

TURKEY, GREECE AND JORDAN

As had by now become customary, Turkey and Greece both received the latest exported American fighter – F-104 deliveries beginning in 1963 and 1964 respectively. Turkey's first 32 F-104Gs came from Lockheed and Canadair production, with the former also supplying four TF-104Gs. They would be joined by 11 more from Spain in 1972: nine F-104Gs and two TF-104Gs. Next, 40 F-104S interceptors were bought from Italy from 1975 to 1976. As Belgium, Canada, Germany, Norway, Germany and the Netherlands phased out their F-104s, more and more of these went to Turkey.

▼ LOCKHEED F-104A STARFIGHTER

Lockheed F-104A Starfighter, 56-879, 9 Squadron, Pakistan Air Force, Sargodha air base, Punjab, Pakistan, 1965. Pakistani Starfighters were extensively used in the conflicts with India, claiming several air-to-air kills.

▼ LOCKHEED TF-104G STARFIGHTER

Lockheed TF-104G Starfighter, 104-03, *Escuadrón 104, Ala 16, Ejército del Aire* (104 Squadron, 16 Wing, Spanish Air Force), Torrejon Air Base, Spain, 1968.
The two-seat variant of the F-104G had a reduced fuel capacity and no gun but still retained some combat capability; three examples of this variant were operated by Spain, alongside 18 single-seat F-104Gs.

Altogether, the Turkish Air Force would receive around 400 Starfighters. Those that lasted long enough would be retired in 1995.

Greece would initially receive 35 Canadian-made F-104Gs and four Lockheed-made TF-104Gs, followed by

AERITALIA F-104S STARFIGHTER

Aeritalia F-104S Starfighter, 9-859, *191 Filo, Türk Hava Kuvvetleri* (191 Squadron, Turkish Air Force), 9th Main Jet Base, Balıkesir, Turkey, 1992.
Turkey was the only export client for the Italian-built F-104S, operating them into the early 1990s.

ten US-made F-104Gs and two more TF-104Gs. Losses resulting from accidents were made up for with nine ex-Spanish F-104Gs in 1972 and two ex-German TF-104Gs in 1977. Ten Italian-made F-104Gs were then received from the Netherlands in 1982. During the 80s, West Germany sent Greece military aid in the form of 38 F-104Gs, 22 RF-104Gs and 20 TF-104Gs – some of them being cannibalised for spares rather than entering Hellenic Air Force service. The last Greek Starfighter was retired in March 1993.

Meanwhile, Jordan had received two F-104As and three F-104Bs from the US in the spring of 1967, with 36 more F-104As and four more F-104Bs following in 1969. They remained in service until 1982-83, when they were replaced by Dassault Mirage F1CJs.

US JET FIGHTERS IN FOREIGN SERVICE 059

US JET FIGHTERS IN FOREIGN SERVICE

VOUGHT F-8 CRUSADER

Difficult to fly but a capable dogfighter and a great bomb truck, the F-8 would serve with just two foreign operators – France and the Philippines.

1964-1999

Designed in response to a 1952 Mach 1.2 carrier fighter requirement, Vought's F-8 Crusader was powered by the Pratt & Whitney J57 and armed with four 20mm cannon in its fuselage sides. Behind the guns was a Sidewinder launch rail – a double rail in later variants. The aircraft had a low undercarriage which also retracted into the fuselage and its wings sat on top of the area-ruled fuselage.

During take-off and landing, the front of the wings would be jacked up seven degrees using hydraulics (or a pneumatic back-up if the hydraulics failed). This meant that while the wings assumed a high angle of attack, the rest of the aircraft remained level, providing the pilot with a better forward field of vision. In addition, a linkage which simultaneously lowered the ailerons and leading edge flaps by 25 degrees decreased take-off and approach speed by providing more lift.

The Crusader entered US Navy service in December 1956 and the French Navy ordered 42 single-seat examples – designated F-8E(FN) – in 1962 for the carriers *Clemenceau* and *Foch*. These aircraft retained their cannon but the French Matra R530 air-to-air missile was usually carried on their rails instead of Sidewinders, requiring the fitment of a Magnavox AN/APQ-104 radar with a modified AN/AWG-4 fire control system. From 1973, the F-8E(RN) could alternatively carry the Matra R550 Magic infrared homing missile, followed by the Magic 2 in 1988.

The first production F-8E(FN) flew on June 26, 1964, and deliveries commenced on October 5, 1964. The French Crusaders got new F-8J wings in 1969 and were re-engined with Pratt & Whitney's J57-P-20A in 1979. Later refurbishment programmes would see 17 examples rewired, strengthened and re-equipped with improved avionics, modified navigation suites and new radar-warning receivers. Starting in 1992, 12 aircraft would receive Martin-Baker zero-zero ejection seats, Mirage F1 navigation systems, a radar altimeter, IFF, ILS and VOR. They were also modified with a Thomson-CSF Sherloc radar warning receiver in a new vertical fin extension. These were finally retired on December 19, 1999, and replaced with Dassault Rafale Ms.

The Philippines bought 35 ex-US Navy F-8Hs in 1977, 25 of them refurbished by Vought and the remainder serving as spare parts. Owing to corruption and mismanagement, the fleet rapidly deteriorated and when Philippine President Ferdinand Marcos left in 1986 the survivors were flown to Clark AFB for their own protection. They would be phased out less than two years later, on January 23, 1988.

▼ VOUGHT F-8E(FN) CRUSADER

Vought F-8E(FN) Crusader, 35, Flottille 12.F, Aeronavale (12F Squadron, French Navy Air Arm), *Clemenceau* aircraft Carrier (R98), 1999.
This aircraft displays the squadron emblem on the tail and carries the French Matra Magic air to-air missile.

▼ VOUGHT F-8H CRUSADER

Vought F-8H Crusader, 48661/301, 7th Tactical Fighter Squadron, *Hukbong Himpapawid ng Pilipinas* (Philippines Air Force), Basa Air Base, Floridablanca, Pampanga, Philippines, 1980.
The Philippine Crusader fleet operated for around a decade before availability issues forced it to be retired.

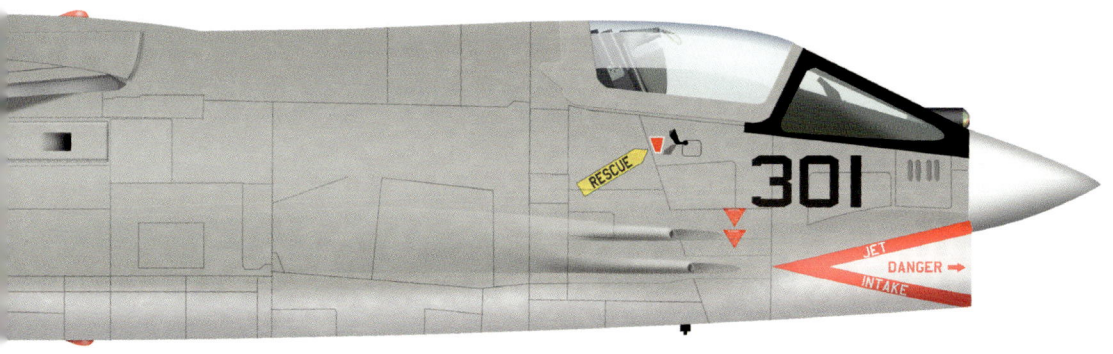

US JET FIGHTERS IN FOREIGN SERVICE

MCDONNELL DOUGLAS F-4 PHANTOM II

The Phantom was perhaps the first truly modern American jet fighter. Despite being a product of the 1950s, heavily upgraded examples continue to see operational service today.

1968–PRESENT

McDonnell began to consider three options for basing a new fighter on the aerodynamic form of its F3H Demon fighter in August 1953. There were three engine options: a Wright J67, two Wright J65s or two General Electric J79s. In each case, the fighter would use a larger folding wing than that of the F3H and the rest of the airframe could be adapted to provide a single- or two-seater layout with equipment for either air-to-air interception, ground-attack or reconnaissance.

An unsolicited proposal was made on September 19, 1953, and the Navy was sufficiently interested in the two twin-engine designs to order a mock-up in early 1954. This was duly produced and inspected on October 18, 1954, resulting in an order from the US Navy for two cannon-armed prototype single-seaters powered by twin J79s. However, this was altered to a pair of missile-armed two-seaters on May 26, 1955, under the designation YF4H-1. Two months later the order was expanded to five pre-production machines.

Armament was four AIM-7s in semi-recessed bays under the fuselage with an option to carry AIM-9s on underwing pylons. The cannon were deleted – making the YF4H-1 America's first missile-only fighter. The wings and engine intakes underwent extensive redesign based on wind tunnel testing and the first prototype Phantom II made its maiden flight on May 27, 1958. After the aircraft won a fly-off against Vought's outlandish-looking XF8U-3 Crusader III in December 1958, the number on order was increased to 45.

After further testing, the aircraft's canopy was revised to improve pilot visibility and a larger radome was installed to house the Westinghouse AN/APQ-72 radar's 32in dish. The intakes now had two ramps ahead of the duct, the first fixed and the second variable, and a retractable air refuelling probe was installed. A Texas Instruments AAA-4 infrared sensor was fitted into a small pod beneath the radome to give the Phantom its now distinctive nose.

The USAF ordered the Phantom as the F-110A or RF-110A in reconnaissance form in January 1962 and eight months later under the common designation scheme the F-110A was renumbered F-4C and the RF-110A became the RF-4C. The original US Navy R4H-1 became the F-4B. The 'C' had full dual controls, more powerful J79s, built-in cartridge

▼ MCDONNELL DOUGLAS F-4E PHANTOM II
McDonnell Douglas F-4E Phantom II, 69-7205/5, 82 Wing, Royal Australian Air Force, RAAF Base Amberley, Queensland, Australia, 1970.
The RAAF flew Phantoms as it awaited the delivery of Australia's F-111s; the aircraft retained their original USAF colour schemes and serial numbers (although RAAF serials were issue to each airframe they were not painted on the aircraft).

▼ MCDONNELL DOUGLAS F-4E PHANTOM II
McDonnell Douglas F-4E Phantom II, 60366/ 7813, 76th Fighter Squadron, 222nd Fighter Regiment, Egyptian Air Force, Fayid Air Base, Egypt, 1980.
Egyptians Phantoms were delivered in South-East Asian colour scheme and later were painted in this new grey finished (called Egyptian One), before adopting the definitive Hill colour scheme (with orange identification markings).

US JET FIGHTERS IN FOREIGN SERVICE — McDonnell Douglas F-4 Phantom II

▼ MCDONNELL DOUGLAS F-4F PHANTOM II

McDonnell Douglas F-4F Phantom II, 37+51, *Jagdgeschwader 71, Luftwaffe* (71 Fighter Wing, German Air Force), Wittmundhafen Air Base, West Germany, 1976.
The Luftwaffe experimented with many colour schemes for its Phantom fleet; this one called *Wolkenmaus* (Cloud Mouse) was not adopted.

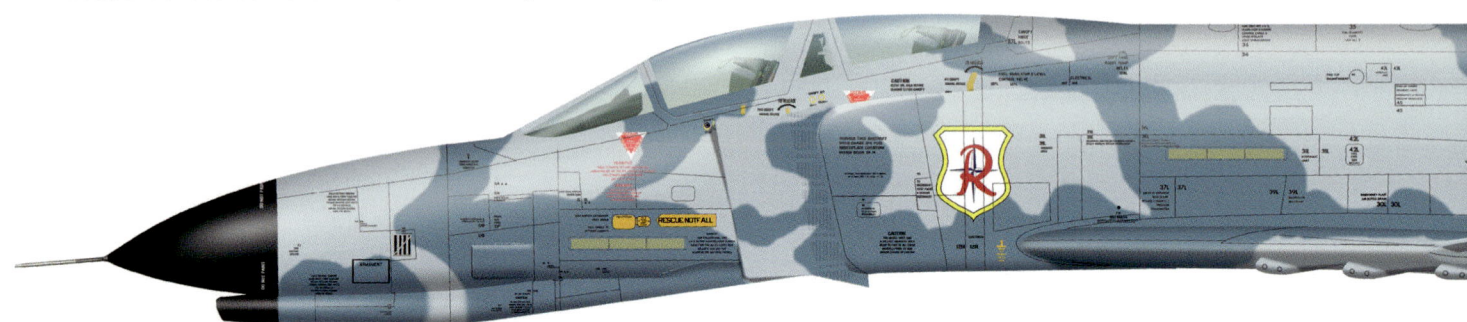

▼ MCDONNELL DOUGLAS F-4E PHANTOM II

McDonnell Douglas F-4E Phantom II, 01517, *337 Mira, Polemikí Aeroporía* (337 Squadron, Hellenic Air Force), Larissa Air Base, Greece, 1992.
Greek Phantoms were used both in air-to-ground and interceptor roles; this aircraft is configured for the latter role, carrying AIM-9 Sidewinder and AIM-7 Sparrow air-to-air missiles.

▼ MCDONNELL DOUGLAS F-4E PHANTOM II

McDonnell Douglas F-4E Phantom II, 3-6528, 61st Tactical Fighter Squadron, Islamic Republic of Iran Air Force, Bushehr Air Base, Iran, 2019.
Phantoms still form a substantial part of the combat force of the IRIAF; this aircraft carries the AGM-65 Maverick air-to-ground missile.

starting system, thicker low pressure tyres for rough field use, a refuelling receptacle in the upper fuselage instead of the naval version's retractable probe and an anti-skid undercarriage braking system. It also had a different electronics suite – the AN/APQ-100 radar, AN/ASN-48 inertial navigation system and AN/ASN-46 navigation computer – and it could carry guided and unguided bombs, AIM-4s, AIM-7s, rockets or gun pods.

The first USAF F-4C flew on May 27, 1963, and topped Mach 2 during the flight, with USAF deliveries beginning in November 1963. In March 1964 the radar was upgraded to the larger but lighter AN/APQ-109A, necessitating a bigger radome, and the navigation system was upgraded. The AN/ASG-22 lead computing sight was fitted and the infra-red sensor was deleted then replaced with a radar sensor and SAM warning receiver – creating the F-4D. The 'D' would continue to be upgraded with additional sensor, targeting and other equipment throughout its long service career – allowing it to carry a wide range of weapons.

Fitting a built-in M61 cannon, deleting the folding wings and once again upgrading to better J79s and a better radar resulted in the F-4E.

Meanwhile, the US Navy had ordered the F-4J – an improved F-4B with J79-GE-10 engines, AN/APG-59 radar, AN/AWG-10 fire control, larger main landing gear wheels, slatted tailplane, drooping ailerons, zero-zero ejection seats and no infrared sensor under the nose.

BRITAIN

The Phantom's first export customer was the UK. The Royal Navy had planned to buying the Hawker Siddeley P.1154 V/STOL fighter for its carriers but after this was cancelled decided instead to purchase a bespoke Phantom variant – the F-4K, essentially an F-4J fitted with powerful Rolls-Royce Spey engines. The British Aircraft Corporation redesigned the F-4J's rear fuselage to accept the Spey and both engines and new fuselage parts were then shipped to McDonnell at St Louis for final assembly. The Spey gave British Phantoms a 10% increase in operational radius, better take-off, better initial climb and better acceleration at low level – but it also offered a reduced top speed and a lower ceiling compared to the J79.

The first F-4K flew on June 27, 1966, and the first examples were delivered in April 1968 and at the end of March 1969, 892 Naval Air Squadron was commissioned as the first Phantom unit. 892 NAS embarked aboard HMS *Ark Royal* with its Phantoms for the first time in June 1970. In British service, the F-4K was designated FG.1 and only 29 were delivered to the Royal Navy.

The last *Ark Royal* Phantom launch was on November 27, 1978, and 892 NAS was disbanded on December 15, 1978. Seven of its Phantoms had been lost in accidents and the survivors were then transferred to the RAF.

With both the TSR.2 and the P.1154 cancelled, the RAF's version of the

US JET FIGHTERS IN FOREIGN SERVICE — McDonnell Douglas F-4 Phantom II

Phantom, designated F-4M, was intended to replace both the Canberra and Hawker Hunter in the reconnaissance, ground-attack and long-range strike roles. In British service it became the FGR.2 and the first of 112 examples arrived at Yeovilton on July 18, 1968. The first unit to receive FGR.2s was 228 Operational Conversion Unit at RAF Coningsby in August 1968, followed by 6 and 54 Squadrons of Air Support Command.

RAF Phantoms were not involved in the Falklands War but did begin operating from Port Stanley Airport in 1982 following its recapture from Argentina. In order to maintain NATO readiness in Europe, Britain bought 15 extensively refurbished ex-US Navy F-4Js under the designation F-4J(UK) or Phantom F.3. These would serve until their retirement in early 1991 with the remaining FGR.2s following in the autumn of 1992.

IRAN

The Shah of Iran ordered 32 F-4Ds in two batches in 1967, the first examples arriving on September 8, 1968. Some of these were used in unsuccessful attempts to intercept Soviet MiG-25 reconnaissance aircraft overflying Iranian territory. One was lost to ground fire during an Iranian operation to help the Sultan of Oman strike rebel positions. And Iranian F-4Ds attacked an Iraqi armoured convoy with AGM-65 Maverick missiles in 1976 during a border clash.

In the early to mid-1970s, the Shah massively expanded his Phantom fleet, ordering a total of 208 F-4Es and 27 RF-4Es. The first examples arrived in March 1971 and 177 F-4Es plus 16 RF-4Es had been delivered by early 1979 when the Shah was forced into exile during an Islamic fundamentalist revolution. All Western military exports to Iran ceased abruptly, leaving Iran with 188 operational Phantoms – some of the earlier aircraft having been lost through accidents and attrition.

Without supplies and spares from the West, these aircraft soon became difficult to operate. When Iran was attacked by Iraq in September 1980, only about 40% of its F-4s were still operational. Those that could fly were initially used to make deep penetration strikes on targets in and around Baghdad but air-to-air combat was rare. Various purges gradually removed experienced Iranian pilots from the military and losses during the first nine months of the war amounted to around 60 Phantoms – many of the remainder being cannibalised for parts.

On June 5, 1984, a Saudi Arabian F-15C Eagle shot down an Iranian F-4E after it strayed too close to Saudi oil facilities. This is the only time when one McDonnell product has been used to shoot down another. By the end of 1986 it was claimed that only 20 F-4Es remained airworthy. However, by acquiring parts from various third parties, Iran has

▼ MCDONNELL-DOUGLAS F-4D PHANTOM II

McDonnell-Douglas F-4D Phantom II, 64-0931, 110th Tactical Fighter Squadron, 11th Fighter Wing, Republic of Korea Air Force, Daegu Air Base, South Korea, 1979.
RoKAF Phantoms were first delivered to the 11th Fighter Wing in 1968.

MCDONNELL DOUGLAS F-4E PHANTOM II

McDonnell Douglas F-4E Phantom II, 332, 119 Squadron, Israel Defense Force- Air Force, Israel, 1973
Delivered during the 1973 conflict, some Phantoms were received and put into immediate operation, still with the original USAF camouflage scheme and painted in Israeli markings.

MCDONNELL-DOUGLAS RF-4C PHANTOM II

McDonnell-Douglas RF-4C Phantom II, 12-67, *Ala 12, Ejército del Aire* (12 Wing, Spanish Air Force), Torrejón de Ardoz Air Base, Spain, 2002.
Modernised Spanish reconnaissance Phantoms soldiered on until the early 2000s, after the fleet of F-4Cs was retired in the early 1990s.

US JET FIGHTERS IN FOREIGN SERVICE — McDonnell Douglas F-4 Phantom II

continued to operate the Phantom. Al Jazeera reported on November 30, 2014, that an Iranian F-4 had been used to attack Islamic State militants in Iraq – later confirmed by the United States.

Today it is believed that Iran still has 64 F-4s – 10 F-4Ds, 50 F-4Es and four RF-4Es – which have been kept going and even upgraded using Chinese-supplied parts.

ISRAEL

The Phantom's largest export customer was Israel – with around 240 F-4Es and RF-4Es being supplied between 1969 and 1976. An initial batch of 44 F-4Es and six RF-4Es was agreed in 1968 under project Peace Echo I. The earliest examples were delivered on September 5, 1969, and flown into battle for the first time on October 22, 1969, when they were used to attack Egyptian SAM sites to the west of the Suez canal. An Israeli F-4 shot down an Egyptian MiG-21 on November 11 and throughout January 1970 Phantoms were used to strike various Egyptian ground targets.

Over the decades that followed, Israeli Phantoms would score a total of 116 victories against Arab aircraft for a loss of 55 F-4s – most to SAMs and AAA. They would be switched to the ground-attack role with the introduction of the F-15 and F-16 during the early 1980s but continued to receive upgrades throughout the 1980s and 1990s. The last examples were retired in 2004.

KOREA

Eighteen F-4Ds were ordered by the Republic of Korea in 1968 and these were

▼ MCDONNELL DOUGLAS F-4E 2020 TERMINATOR PHANTOM II

McDonnell-Douglas F-4E 2020 Terminator Phantom II, 77-0300, *401 Filo, Türk Hava Kuvvetleri* (401 Squadron, Turkish Air Force), 1st Main Jet Base Command, Eskisehir, Turkey, 2019. The *401 Filo* is a test unit, tasked with several weapons' testing; this aircraft carries a bomb with the HGK guidance kit (GPS/INS).

▼ MCDONNELL DOUGLAS F-4K FG.1 PHANTOM II

McDonnell Douglas F-4K FG.1 Phantom II, XT866/VL-158, 767 Naval Air Squadron, Fleet Air Arm, RNAS Yeovilton, 1971. The Royal Navy's Phantoms operated from aircraft carriers and were shore-based at Yeovilton.

transferred over from ex-USAF stocks under the Peace Spectator programme. The first examples arrived in August 1969. Another 18 were received in 1972. The Republic of Korea Air Force next ordered 37 new F-4Es and these were delivered starting in 1978 under Operation Peace Pheasant II. Six more F-4Ds were delivered in 1982, replacing losses due to attrition, then 24 more were delivered between December 1987 and April 1988. Twenty-three ex-USAF RF-4Cs were received starting in 1990.

In total, the South Koreans would receive 92 F-4Ds, 23 RF-4Cs and 103 F-4Es, of which 19 F-4Es remain in service today.

AUSTRALIA

Having ordered the General Dynamics F-111C to replace its aging fleet of Canberra light bombers, Australia was forced to borrow 24 new F-4Es from the US under the Peace Reef programme in 1970 when it became clear that the F-111 would not be ready for some time to come. One of the F-4Es was lost in an accident on June 16, 1971, during night bombing practice and the remaining 23 had been returned to the USAF by June 21, 1973.

GERMANY

West Germany ordered a total of 88 RF-4Es – an export-only variant which never saw USAF service – in January 1969 and the type entered Luftwaffe service on January 20, 1971. That same year Germany ordered 175 F-4Fs, the F-4F being a lighter, simpler and cheaper version of the F-4E incorporating a number of German-made major components. The first F-4F flew on May 18, 1973, and deliveries took place from September 5, 1973, to April 1976.

Starting in 1978, the RF-4Es were upgraded with ground-attack capability by Messerschmitt-Bölkow-Blohm. They received under-wing hardpoints and the associated wiring as well as a bomb aiming sight and weapon selection switches in both front and rear cockpits.

US JET FIGHTERS IN FOREIGN SERVICE

McDonnell Douglas F-4 Phantom II

▼ MCDONNELL DOUGLAS F-4EJ PHANTOM II

McDonnell Douglas F-4EJ Phantom II, 27-8306, 306 Squadron, Japanese Air Self Defence Force, Komatsu air base, Japan, 1986. A special colour scheme to celebrate the squadron's 25th anniversary; the last three digits of the serial may have been a factor in this aircraft being chosen to carry the colour scheme.

And between November 1980 and 1983 the F-4Fs were retrofitted with inflight refuelling receptacles, a digital weapons computer, improved electronic countermeasures equipment, cockpit displays, all-weather systems and the ability to fire Sparrow missiles, which had been initially excluded from the design on the grounds of cost, plus the AGM-65 Maverick and the AIM-9L Sidewinder variant. Radar upgrades then followed during the 80s and early 90s.

All the RF-4Es were retired in 1993-94 and replaced with Tornados while the last F-4Fs were retired on June 29, 2013.

JAPAN

A new version of the Phantom was designed specifically for Japan's needs as the F-4EJ. It was intended solely for interception, with no ground-attack capability, and it was decided that it should be manufactured under licence in Japan by Mitsubishi Heavy Industries.

▼ MCDONNELL DOUGLAS F-4J PHANTOM II

McDonnell Douglas F-4J Phantom II, XV582/M, 111 Squadron, Royal Air Force, RAF Museum Cosford, UK, 2017.
The appropriately named Black Mike was made famous for its record flight between John O'Groats and Land's End in 1988; the aircraft is currently preserved at Cosford.

The order was agreed on November 1, 1968, and the first two F-4EJs were built by McDonnell in St Louis and flight testing began on January 14, 1971. Another 11, built in kit form, were assembled in Japan by Mitsubishi. The first all-Japanese example made its flight debut on May 12, 1972. Another 127 were then made with the last being delivered on May 20, 1981 – the last new Phantom to be built anywhere.

From November 1974 to June 1975, 14 McDonnell-built RF-4EJs were delivered. These were nearly identical to the American RF-4C. Numerous upgrades would follow until the RF-4Es were retired in May 2020, followed by the last F-4EJs in March the following year.

SPAIN

A batch of 36 ex-USAF F-4Cs was acquired by Spain between October 1971 and September 1972 under the Peace Alfa programme. The Spanish gave them the designation C.12. Four more F-4Cs plus four RF-4Cs were received under Peace Alfa II in October 1978, the latter being given the designation CR.12. All these had been withdrawn from front-line service by April 1989. Then eight more RF-4Cs, updated with new equipment and new engines, were delivered that year. Another six followed in October 1995. This second set of Phantoms were retired in 2002.

TURKEY

As with many earlier American types, Turkey was a significant operator of the Phantom – acquiring a total of 233 F-4Es and RF-4Es. The first 40 brand new F-4Es were delivered under project Peace Diamond I starting in August 1974. Peace Diamond III saw Turkey receive another 32 new F-4Es and eight RF-4Es in 1977, followed by 15 ex-USAF F-4Es from June 1981 and yet another 15 in mid-1984. Then Peace Diamond IV saw 40 more delivered from June to October 1987 and another 40 were sent as payment for US use of Turkish bases during the Gulf War, starting in March 1991. And from 1992 to 1994, Turkey took delivery of 32 ex-Luftwaffe RF-4Es.

Fifty-four F-4Es were upgraded to Phantom 2000 standard by Israel Aircraft Industries starting in 1997, receiving new radar equipment, air data computer, electronic warfare systems, GPS and hand-on throttle and stick. These aircraft are unofficially known as F-4E 2020 Terminators and 48 remain in service today.

GREECE

The Peace Icarus project saw Greece receive 36 F-4Es, starting in March 1974, with two further aircraft arriving in June 1976. Eight RF-4Es plus another 18 F-4Es followed in 1978-79. Then 28 more F-4Es were delivered in late 1991. Greece also received 29 ex-Luftwaffe RF-4Es in 1993. A series of upgrades followed and today the Hellenic Air Force continues to operate 35 modernised F-4Es.

EGYPT

As a result of the Camp David agreement, the US supplied 35 ex-USAF F-4Es to Egypt under the Peace Pharoah project in 1979. Egyptian pilots found them significantly harder to fly than their old MiG-21s and ground crews initially found them difficult to maintain. The US provided training programmes during the early 80s however and the situation improved. Another seven F-4Es were delivered in 1988 and the fleet continued to fly, subject to various updates and strengthening work, until their retirement in 2020.

US JET FIGHTERS IN FOREIGN SERVICE

NORTHROP F-5
FREEDOM FIGHTER/TIGER II

Though it would enjoy a remarkable career in T-38 Talon trainer form, Northrop's F-5 fighter would only be operated in small numbers by the USAF and US Navy. On the export market, however, it was another smash hit – serving with more foreign air forces than any other American jet fighter.

1965–PRESENT

The N-102 Fang was Northrop's entry for the USAF's WA-303A weapon system requirement in early 1953. It was a small design with a high delta wing, either a V-tail or single conventional fin and a semi-circular under-fuselage inlet for its single J79 or Sapphire engine.

Though defeated by what would become the F-104, Northrop revisited the design in 1954 after NATO issued a requirement for an easy-to-maintain lightweight tactical strike fighter capable of carrying both conventional and tactical nuclear weapons from rough airfields.

Rather than evolving the Fang in isolation however, Northrop decided to visit potential customer nations both in NATO and SEATO to see what their requirements actually were. These turned out, unsurprisingly, to be high performance and manoeuvrability – but also high reliability, ease of maintenance, simplicity of design and service longevity.

By March 1955, Northrop had decided to use two lightweight J85 engines, initially placing them under the aircraft's wings to create the N-156TX. This design also had a more conventional layout with tailplanes and a single fin. The finalists for the NATO requirement were announced in June 1955 – with Northop missing out – but the company pressed ahead using its own funds, convinced that the design had merit.

By November 1955, it had evolved into the low-wing T-tail PD-2706, which had its two engines side by side within its fuselage, fed by side inlets. The evolutionary process continued throughout 1956, with the refined low-wing, low-tailplanes, single-seat N-156F fighter configuration emerging that October alongside the two-seater N-156T advanced trainer.

The latter was chosen to replace the T-33 in July 1956 but the N-156F had to wait until February 1958 before an order was received – for three N-156F prototypes with the intent that the resulting F-5A would be a low-cost fighter for export. The type featured an area-ruled fuselage and viewed in profile it was slightly bent to provide a better view from the cockpit. Armament was two 20mm M39 cannon in the upper nose plus an AIM-9 Sidewinder on each wingtip. Stores could be carried on two underwing stations and a centreline pylon.

Testing was favourable but it would be more than four years before the type was finally ordered in quantity, alongside the F-5B two-seat trainer, in October 1962. Deliveries to foreign nations commenced in 1965.

During 1969, Northrop began work on a much-improved F-5 powered by the new J85-GE-21A. The more powerful engines required enlarged air intakes and the fuselage was lengthened by 15in and widened by 16in. This also provided more space for fuel. The wingroot leading-edge extensions were also enlarged, creating greater wing area, and a system of manoeuvre flaps was introduced. All of this resulted in the modified type, initially dubbed F-5A-21, having a 23% better rate of climb, a 39% better turn radius

▼ CANADAIR CF-5A
Canadair CF-5A, OJ-1, Z28 Fighter Squadron, Botswana Defence Force Air Wing, Maparangwane Air Base, Molepolole, Botswana, 2002.
Botswana has operated a squadron of CF-5s from the mid-1990s on; there are reports of a programme to replace these fighters, although no confirmation on the type yet.

▼ NORTHROP F-5EM TIGER II
Northrop F-5EM Tiger II, 4857, 1º Esquadrão, 14º Grupo de Aviação "Pampa", Força Aérea Brasileira (1st squadron, 14th aviation Group, Brazilian Air Force), Canoas air base, Rio Grande do Sul, Brazil, 2021.
The FAB has operated the F-5 for almost 50 years and upgraded them to F-5EM standard in 2011; this aircraft carries Rafael Python 4 and Derby air-to-air missiles.

US JET FIGHTERS IN FOREIGN SERVICE — Northrop F-5 Freedom Fighter/Tiger II

and a maximum speed increase from Mach 1.4 to Mach 1.6.

The F-5A-21 also featured an AN/APQ-159 lightweight miniature X-band pulse radar for air search and range tracking plus an AN/ASG-31 lead-computing optical gunsight. In December 1970 the aircraft was ordered into production as the F-5E Tiger II, with the first being rolled out in June 1972. The combat trainer version was the F-5F, which had a lengthened fuselage but only one 20mm cannon.

IRAN

The F-5A Freedom Fighter's first overseas recipient was Iran, with 11 F-5As and two F-5Bs being delivered in February 1965. Overall, Iran received 104 F-5As and 23 F-5Bs. The Shah then ordered the new F-5E and F-5F in 1972, with a combined total of 166 being delivered between 1974 and 1976. The F-5As and Bs had been sold off in the meantime.

Following the Islamic revolution, spare parts became difficult to obtain and by 1983 only 40-65 F-5s remained airworthy, with that number reportedly dipping as low as 10-15 by 1986. Iranian F-5s served throughout the Iran-Iraq war, scoring victories but also suffering losses. After the war, Iran was able to refurbish and restore a number of F-5s and up to

▼ NORTHROP F-5E TIGER II

Northrop F-5E Tiger II, 4007, *Escuadrilla de Caza, Fuerza Aérea Hondureña* (Fighter Squadron, Honduran Air Force), Coronel Hector Caraccioli Moncada air base, La Ceiba, Honduras, 2019.
Controversial at the time of their delivery, Honduran F-5Es were introduced as a mean of counteracting the perceived threat posed by MiG-21s from neighbouring Nicaragua; the aircraft still provide Honduras with a fast jet fighter capability today.

▼ NORTHROP F-5F TIGER II

Northrop F-5F Tiger II, 3-7155, 43rd Tactical Fighter Squadron, Islamic Republic of Iran Air Force, 4th Tactical Air base, *Vahdati*, Iran, 2017.
After operating the F-5 for nearly 50 years, including extensive operational use during the Iran-Iraq war, Iran continues to field several squadrons of this fighter jet.

40 reportedly remain in service today – including Iranian copies made by HESA.

GREECE
Starting in 1965, the Hellenic Air Force received around 70 F-5s including F-5As, Bs and RF-5As – the reconnaissance variant. Then another 10 were bought from Iran in 1975 along with 10 from Jordan. Yet another 10 were acquired from Norway in 1986 and five years later 10 NF-5As were purchased from the Netherlands. The last Greek F-5s were withdrawn in 2002.

SOUTH KOREA
The first F-5As arrived in Korea during early April 1965 – with the total eventually reaching 88 F-5As, 30 F-5Bs and eight RF-5As. In 1972, all of the RF-5As and 36 F-5As were transferred to the South Vietnamese Air Force as a favour to the US government.
Nineteen escaping South Vietnamese F-5Es were absorbed into the RoKAF in 1974 and Korea would go on to acquire 233 F-5E/Fs up to 1986. The Hanjin Corporation built the last 68 of these under licence in Korea. Today all the Freedom Fighters have long since been retired but 80 Tiger IIs remain in RoKAF service.

NORWAY
Norway's first Freedom Fighters were delivered in June 1965 and it would eventually receive 78 F-5As, 16 RF-5As and 14 F-5Bs. By the mid-1970s 16 of these aircraft had been wrecked in accidents. Cracks discovered in the engine air intake ducts resulted in the survivors being grounded in 1982. Updates and repairs followed and the Norwegian F-5 fleet was eventually retired in August 2000.

TURKEY
A total of 75 F-5As, 13 F-5Bs and 20 RF-5As were received by Turkey from 1965 to 1972. Following the suspension of US aid in 1974, Turkey got another six F-5As and one F-5B from Libya. Then from 1983 to 1987 yet another 26 F-5As and six RF-5As were received from Norway. The Netherlands sent 44 F-5As and 16 NF-5Bs from 1989 to 1991. The last 11 NF-5s were retired in 2013.

TAIWAN
The first of 72 F-5As and 11 F-5Bs arrived in Taiwan during 1965. In 1972, 48 F-5As were loaned to South Vietnam but a year later most of those were in a non-flyable condition. The US therefore repaired and returned 11 of them plus nine from US reserves. Then 28 new F-5Es were received up to May 1975 to make up the total of 48. The Taiwanese Aero Industry Development Center then commenced local F-5E and F production. Eventually the Republic of China Air Force had received 242 F-5Es and 66 F-5Fs. At the time of writing, the RoCAF continued to operate 27 F-5E/RF-5Es with plans to retire them in 2026.

THAILAND
In exchange for the use of its air bases, Thailand began to receive F-5s in 1966. It would eventually receive 24 F-5As, four RF-5As and two F-5Bs. These were followed by 17 F-5Es and three F-5Fs in 1977, then another 17 F-5Es and three F-5Fs in 1981. Then 10 more F-5Es were received in 1988. Today, following various modifications and modernisations, the Royal Thai Air Force continues to operate 34 F-5E/Fs.

VIETNAM
The USAF deployed F-5As to South Vietnam under the Skoshi Tiger programme in July 1965 and the following year deliveries to the Republic of Vietnam Air Force commenced. By 1973 the VNAF had received a total of 126 F-5As – many former South Korean, Iranian or Taiwanese airframes. At the point when

US JET FIGHTERS IN FOREIGN SERVICE
Northrop F-5 Freedom Fighter/Tiger II

Saigon was overrun by North Vietnamese forces, the VNAF had 82 operational F-5As/RF-5As with another 36 in storage. Some of the airworthy examples were flown to Thailand by their escaping pilots but North Vietnam captured 87. Some of these were then operated by North Vietnam, some were sold internationally and a handful were turned over to the Soviets for examination. The airworthy examples were eventually rendered unserviceable due to lack of parts but a number reportedly remain in storage.

ETHIOPIA
At least 12 F-5As and two F-5Bs were received by Ethiopia in 1966, followed by eight F-5Es in 1975. During a conflict with Somalia in 1977, Ethiopian F-5Es gained air superiority thanks to their AIM-9 Sidewinders – though at least three were shot down by ground fire. During the late 1970s, Ethiopia received a handful of F-5A and E fighters from Vietnam but all of Ethiopia's F-5s are believed to have been transferred to Iran in 1986.

PHILIPPINES
Hard lobbying from President Ferdinand Marcos resulted in 19 F-5As and three F-5Bs being delivered to the Philippines in October 1966. These would later be supplemented by another 15 sourced from Taiwan and South Korea. By the early 1990s, following the end of Marcos's dictatorial rule, eight F-5As and two F-5Bs remained and the survivors were decommissioned in 2005.

MOROCCO
From 1966 to 1974 the Royal Moroccan Air Force received 22 F-5As, two F-5Bs and two RF-5As from the US. Three were involved in a failed coup attempt in 1972 – trying to shoot down King Hussain II's Boeing 727 then strafing and bombing both the royal palace and a military airfield. Their pilots were later arrested.

Two more F-5As were received from Iran in 1974, then another six from Jordan in 1976. During the Western Sahara War, Moroccan F-5s flew bombing and reconnaissance missions – several being lost to ground fire. Sixteen F-5Es and four F-5Fs were received from 1981 to 1983, then 12 more F-5Es starting in 1990. At the time of writing Morocco still operated 24 F-5Es plus four F-5Fs for training.

CANADA
The Canadian government agreed to licence production of the F-5 in 1965 with Canadair building the airframes and Orenda the engines. Initially 115 CF-5As (the official designation, confusingly, was CF-116) were ordered with another 105 to be constructed for the Netherlands as the NF-5. In addition to Orenda modifying the J85 for improved output, the Canadian F-5s also featured two-position nose gear, which reduced the take-off run by

▼ NORTHROP F-5E TIGER II
Northrop F-5E Tiger II, 903, Fighter Squadron, Kenyan Air Force, Laikipia air base, Nanyuki, Kenya, 1980.
Kenya has used its Tigers in several border and internal conflicts since its delivery in 1978.

▼ NORTHROP F-5E TIGER II
Northrop F-5E Tiger II, J-3082, *Patrouille Suisse* demonstration team, *Schweizer Luftwaffe/ Forces Aériennes Suisses/ Forze Aeree Svizzere* (Swiss Air Force), Emmen Air Base, Lucerne, Switzerland, 2014.
Patrouille Suisse started flying the F-5 in 1994 and still operates them today; this aircraft carries markings celebrating the 50th anniversary of the team in 2014.

NORTHROP F-5E TIGER II

Northrop F-5E Tiger II, 4509, *Escuadrón Aéreo 401, Fuerza Aérea Mexicana* (Air Squadron 401, Mexican Air Force), Santa Lucia air base, México City, Mexico, 2007.
F-5s have provided Mexico with a jet fighter force since the early 1980s; this aircraft carries tail art celebrating 25 years of operation in 2007.

US JET FIGHTERS IN FOREIGN SERVICE
Northrop F-5 Freedom Fighter/Tiger II

▼ NORTHROP F-5F TIGER II

Northrop F-5F Tiger II, 5403/30129, 7th Flight Training Wing, Republic of China Air Force, Songshan air base, Taipei, Taiwan, 2018.
This F-5F carries tail art celebrating the 40th anniversary of the 7th Flight Training Wing.

▼ NORTHROP F-5TH SUPER TIGRIS

Northrop F-5TH Super Tigris, 41575/21137, 211 Squadron, 21 Wing, Royal Thai Air Force, Ubon Air Base, *Ubon*-Ratchathani, Thailand, 2019.
The Super Tigris upgrade programme saw the modernisation of the RTAF's F-5Es and F-5Fs, with improvements to the structure, avionics and weapons capabilities; this aircraft carries IRIS-T and Rafael Derby air-to-air-missiles.

▼ NORTHROP F-5E TIGER II

Northrop F-5E Tiger II, Y-92509, *Escadron 15,* (Squadron 15), Tunisian Air Force, Bizerte-Sidi Ahmed air base, Tunisia, 2001.
Delivered in 1984, Tunisian F-5s are being considered for replacement, although still no firm confirmation is known.

20% by increasing the angle of attack.

The first CF-5A was rolled out in 1968 with the first NF-5 following in 1969. Eventually Canadair would build 89 single seat CF-5As, 46 two-seater CF-5Ds, 75 single seat NF-5As and 30 two-seater NF-5Bs. Upgrades were made to the CF-5s up to 1995 when the fleet was retired.

NETHERLANDS

Dutch plans to co-produce more than 200 F-5s in collaboration with Belgium had to be scrapped when Belgium decided to buy the Mirage 5 instead. Therefore, in 1967, the Dutch government decided to buy F-5s from the Canadian production line instead – since the Canadian modifications suited the requirements of the Royal Netherlands Air Force. As mentioned, Canadair built 75 NF-5As and 30 NF-5Bs for the Netherlands with deliveries from 1969 to 1972. The last NF-5s were withdrawn in 1991.

SPAIN

Licence production of the F-5 also took place in Spain, with Construcciones Aeronauticas S.A. (CASA) building 18 SF-5As, 34 SF-5Bs and 18 SRF-5As. Deliveries took place from 1968 to 1971. Surviving aircraft underwent a programme of upgrades starting in 1988. Today Spain retains 19 two-seat F-5s for use as jet trainers – SF-5Bs upgraded to F-5M standard by Israel Aircraft Industries – with the others having been retired.

LIBYA

As an independent state under the rule of King Idris I, Libya was a recipient of American aid – eight F-5As and two F-5Bs being delivered in 1968. The following year a military coup deposed the king, bringing Colonel Muammar al-Gaddafi to power and ending US aid. While the F-5s were reportedly kept flying by Greek personnel, spares eventually dried up and seven were handed to Turkey in 1975. The fate of the remainder is unclear.

BRAZIL

Thirty-six F-5Es and six F-5Bs were ordered by Brazil in 1974. The aircraft were delivered and served into the 1980s. They were supplemented by 17 refurbished F-5Es and three F-5Fs under the Peace Amazon II project in 1989. A series of upgrades followed, with the F-5Es being redesignated F-5EM and the F-5Fs becoming F-5FMs. and the Brazilian Air Force currently operates 42 F-5EMs plus four F-5FMs.

SAUDI ARABIA

Between 1974 an 1985 Saudi Arabia received 20 F-5Bs, 109 F-5E/Fs and 10 RF-5Es under the Peace Hawk project. One Saudi F-5 was destroyed by ground fire during the Gulf War in 1991 while flying close air support missions against Iraqi units in Kuwait. Its remaining aircraft were offered for sale from 2009 to 2014.

US JET FIGHTERS IN FOREIGN SERVICE — Northrop F-5 Freedom Fighter/Tiger II

VENEZUELA
Sixteen of Canada's CF-5As and two CF-5Ds were sold to Venezuela in 1972, these being given the local designations VF-5A and VF-5D. Two of the VF-5As were converted into reconnaissance variants as RVF-5As. Seven had been wrecked by 1990 and the remainder were then put into storage due to metal fatigue problems. Later that year Venezuela received six ex-Dutch NF-5Bs and one NF-5A. During a failed coup against the Venezuelan government in 1992, three VF-5As were destroyed on the ground and one was scrambled to defend Barquisimeto Air Base against rebel forces. All F-5s had been withdrawn from Venezuelan service by 2015.

JORDAN
When Iran disposed of its F-5A/Bs in 1974-1975, Jordan acquired 30 As and six Bs. The US then directly supplied 61 F-5Es and 12 F-5Fs in 1975. Most of Jordan's F-5s were sold off during the 1990s, 2000s and 2010s.

CHILE
The military junta of General Augusto Pinochet received 15 F-5Es and three F-5Fs in 1976 under the Nixon administration. The Carter administration, however, put an arms embargo to Chile in place due to Pinochet's human rights record and just four F-5s were still flying by the mid-80s due to a lack of spares. Relations with the US improved after Pinochet left office and IAI upgraded 12 of Chile's F-5Es and two F-5Fs to Tiger III status. Today Chile still operates 10 F-5Es.

MALAYSIA
Fourteen F-5Es and two F-5Bs were received by Malaysia in 1975, with the latter two aircraft being transferred to Thailand in 1982 when four additional F-5Fs were received. The following year they were joined by two RF-5Es. Five F-5Es and one F-5F were lost in accidents and the survivors were retired in 2015.

SWITZERLAND
Under the 1976 Peace Alps project, 66 F-5Es and six F-5Fs were ordered by Switzerland. The first 13 F-5Es and all of the F-5Fs were delivered from the US with the remaining 53 F-5Es being assembled by FFA in Switzerland. Another 32 F-5Es and six F-5Fs were ordered in 1981 – the first F-5E coming from Northrop and all the rest being assembled by FFA. Upgrades followed and today Switzerland still operates 27 F-5Es.

KENYA
Ten F-5Es and two F-5Fs were delivered to Kenya in 1978 with another two F-5Fs following in 1982 as attrition replacements. Eventually, Kenya bought

▼ NORTHROP F-5A FREEDOM FIGHTER
Northrop F-5A Freedom Fighter, FA-495, 10th Fighter Wing, Republic of Korea Air Force, Suwon Air Base, South Korea, 1970. South Korea operated several squadrons of the early F-5A/Bs and still operates the later F-5E/Fs; during the Vietnam War, some aircraft were transferred to South Vietnam in exchange for the delivery of F-4Ds from the US.

▼ NORTHROP F-5A FREEDOM FIGHTER
Northrop F-5A Freedom Fighter, 10271, *Khong Quan Nhan Dan Viet Nam* (Vietnam People's Air Force), Bien Hoa Air Base, Đồng Nai, Vietnam, 1980.
With the fall of Saigon, several aircraft were inducted into service with the Vietnam People's Air Force, including many F-5s; the aircraft were operated until the early 1980s.

▼ NORTHROP F-5A FREEDOM FIGHTER
Northrop F-5A Freedom Fighter, 13318, 522nd Fighter Squadron, *Tho Quoc Khong Quân (*Republic of Vietnam Air Force), Da Nang air base, South Vietnam, 1968.
South Vietnam was a recipient for both the F-5A and latter F-5E aircraft; the aircraft were heavily involved in operations right until the end of the conflict.

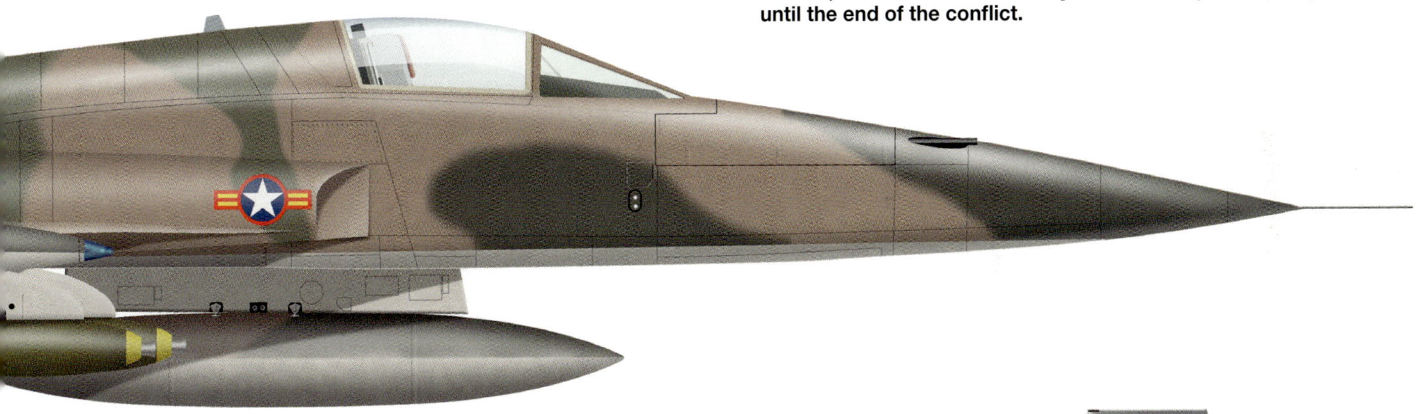

US JET FIGHTERS IN FOREIGN SERVICE
Northrop F-5 Freedom Fighter/Tiger II

▼ NORTHROP F-5E TIGER II

Northrop F-5E Tiger II, TS-0510, *Skadron Udara 14, Tentara Nasional Indonesia Angkatan Udara* (14th Squadron, Indonesian Air Force), Iswahyudi air base, East Java, Indonesia, 1986. Received in 1980, Indonesian F-5s have been put into reserve/storage, awaiting replacement.

▼ NORTHROP F-5E TIGER II

Northrop F-5E Tiger II, 91924, *Escadre de Chasse 2* (2nd fighter squadron), Royal Moroccan Air Force, Bassatine air base, Meknes, Morocco, 2016.
Moroccan F-5s were used operationally during the conflict in West Sahara; the fleet was upgraded in 2001, improving the avionics and weapons capabilities.

another 10 F-5Es and 23 F-5Fs from Jordan – these having been upgraded to F-5EM standard. Today the Kenyan Air Force maintains 17 F-5EMs and six F-5F trainers.

YEMEN
The Saudi government transferred four F-5Bs to North Yemen in 1979 with US approval, so they could be used to counter South Yemen's MiGs. A handful of F-5Es followed but all were unserviceable by 1994. More supplies of spares were forthcoming from the US in 1995 and today the Yemen Air Force has 13 serviceable F-5s.

SINGAPORE
Singapore received 18 F-5Es and three F-5Fs in 1979, with another six F-5Es and three additional F-5Fs arriving between 1981 and 1983. Yet another six F-5Es were received in 1986, three more F-5Fs in 1987, and five more F-5Es in 1989 – built from spares, since the F-5 production line had ended by this point. Eight F-5Es were converted to RF-5E status in 1990 and seven more F-5Es were acquired from Jordan in 1994. All had been retired by 2010.

INDONESIA
Eight F-5Es and four F-5Fs were acquired by Indonesia in 1980 with four more F-5Es being received later. All 16 were retired by 2005 and remain in storage today.

SUDAN
Ten F-5Es and two F-5Fs were delivered to Sudan from 1982 to 1984, with one F-5F later being sold to Jordan. Two F-5 pilots defected to Sudan from Ethiopia, bringing their aircraft with them. American economic assistance to Sudan ended in 1989 with the F-5s believed to have become unserviceable not long after.

MEXICO
Ten F-5Es and two F-5Fs were received between August and November 1984. One F-5E was wrecked in a mid-air collision during a parade flight on September 16, 1995. The remainder were retired in 2017 but two years later four F-5Es and one F-5F were put back into service after repaired engines were received.

TUNISIA
Eight F-5Es and four F-5Fs were sent to Tunisia between 1984 and 1985, plus another five ex-USAF F-5Es in 1989. Today 11 F-5Es and three F-5Fs remain in Tunisian service.

HONDURAS
Honduras received the now seemingly standard export package of 10 F-5Es and two F-5Fs in 1987. Today the Honduran Air Force still flies three F-5Es and two F-5Fs.

BAHRAIN
The Bahrain Amiri Air Force took delivery of eight F-5Es and four F-5Fs in 1988. All 12 still fly with the renamed Royal Bahraini Air Force – but only as conversion trainers for the force's F-16s.

BOTSWANA
Ten upgraded CF-5As and three CF-5Ds were bought from Canada by Botswana in 1996, with another three CF-5As and two CF-5Ds following in 2000. The Botswana Defence Force Air Wing flies 11 CF-5As and four CF-5Ds.

AUSTRIA
A dozen F-5Es were leased from Switzerland in 2005 as a stopgap measure pending the delivery of Austria's Eurofighter Typhoons. When these were received, the F-5Es were returned.

▼ **CANADAIR CF-116A(R) FREEDOM FIGHTER**
Canadair CF-116A(R) Freedom Fighter, 726, No. 433 Escadrille Tactique de Combat (No 433 Tactical Combat Squadron), Royal Canadian Air Force/ Aviation Royale Canadienne), Canadian Forces Base Bagotville, Quebec, Canada, 1972. This aircraft is seen here with the optional reconnaissance nose section.

US JET FIGHTERS IN FOREIGN SERVICE

GRUMMAN F-14A TOMCAT

Grumman's superb variable geometry fleet defence fighter infamously only ever had one export customer – Iran.

1976–PRESENT

▼ GRUMMAN F-14A TOMCAT

Grumman F-14A Tomcat, 3-863, Imperial Iranian Air Force, Tactical Air Base 8, Isfahan, Iran, 1977.
The first Tomcat for Iran is seen carrying the full panoply of air-to-air missiles: AIM-54 Phoenix, AIM-7 Sparrow and AIM-9 Sidewinder.

▼ GRUMMAN F-14A TOMCAT

Grumman F-14A Tomcat, 3-6041, Islamic Republic of Iran Air Force, Iran, 2021.
Iranian Tomcat are still flying and their full operational status remains a matter of great debate; this F-14A was seen carrying a Russian R-27 air-to-air missile.

When the swing-wing F-111B was cancelled in 1968, the US Navy launched the VFX programme for a tandem two-seat, twin-engine, Mach 2.2 air-to-air fighter. Grumman won the contest with its 303E design in January 1969 and this would become the F-14.

It was a tandem two-seater with a broad flat fuselage and variable sweep wings attached to non-moving 'gloves' which housed their rotation mechanism. It was initially powered by two Pratt & Whitney TF30-P-412 engines in under-slung nacelles – though this was soon replaced by the TF30-P-414 due to reliability issues. Armament was one M61 cannon under the left hand side of the cockpit plus six AIM-54 Phoenix missiles or four fuselage AIM-54s plus two AIM-7s and two AIM-9s.

The AIM-54s and AIM-7s were linked to a Hughes AN/AWG-9 radar with AN/AWG-15 fire control system, managed by the radar intercept officer in the back seat. Under its nose, the F-14A had an AN/ALR-23 Infrared Search and Track sensor. Other key systems included AN/ALE-39 chaff and flare dispensers, an AN/APR-45 radar warning receiver and an AN/ALQ-126 electronic jamming system.

Structurally, the aircraft was primarily made from lightweight aluminium alloy but around 25% was titanium, including the wing box, wing pivots, upper and lower wing skins. The type entered US Navy service in September 1974.

Before then, however, when the Shah of Iran wanted an aircraft to intercept MiG-25 Foxbat overflights of his territory in May 1972, US President Richard Nixon offered him two options – the F-14 or the F-15. The Shah chose the former and Iran became the Tomcat's only export customer.

An order for 30 was initially placed in January 1974 with 50 more being added five months later. The Iranian F-14As were basically the same as those of the US Navy with the omission of some classified electronics. They had the improved TF30-P-414 engine and the first examples were delivered in January 1976. The last of 79 arrived in 1978 with one being retained by the US for use in experiments. A total of 714 AIM-54A Phoenix missiles had also been ordered but only 284 had been delivered by the time of the Islamic Revolution – which abruptly halted all US exports to Iran.

During the Iran-Iraq War of 1980-1988, availability of Iranian Tomcats was low due to lack of parts but they were sometimes used as mini-AWACS due to their powerful radar units. The Iraqis claimed to have shot down at least ten F-14As while the Iranians claimed their Tomcats had shot down around 35-40 Iraqi fighters for only one loss in air-to-air combat. Recent research suggests that Iranian Tomcat kills might in fact have been as high as 100.

The Iranians have evidently been able to keep around 30 F-14As operational at any one time and still have a total of around 50-55 today.

US JET FIGHTERS IN FOREIGN SERVICE

MCDONNELL DOUGLAS/BOEING F-15 EAGLE

For pure air superiority the F-15 has so far proven to be unmatched and newer variants have offered superior ground-attack capability too. For decades there were famously just three foreign operators – Israel, Japan and Saudi Arabia. But three more have recently adopted the Eagle.

1976–PRESENT

▼ MCDONNELL DOUGLAS F-15I RA'AM
McDonnell Douglas F-15I *Ra'am*, 234, 69 Squadron, Israel Defense Force – Air & Space Force, Hatzerim Airbase, Israel, 2014.
The F-15I is operated by a single squadron within the IAF; this aircraft carries the indigenous Rafael Python 3 air-to-air missiles.

During the late 1960s it became clear that the earlier shift towards high-speed, missile-only fighters had been a step in the wrong direction. Combat experience in Vietnam indicated that a lighter, slower, cheaper but much more manoeuvrable fighter, armed with both guns and missiles, was needed.

By December 1968 the finalists to produce America's next fighter had been narrowed down to McDonnell Douglas, Fairchild Republic and North American Rockwell. A year later, McDonnell Douglas's design emerged victorious as the F-15. It had a flat broad fuselage, fixed wings, twin-fins and big shoulder intakes feeding its two Pratt & Whitney F100 engines – with an M61 Vulcan 20mm cannon fitted as standard.

It had two underwing hardpoints, each able to take a pair of missile launch rails, four hardpoints on the underside of its fuselage for semi-recessed AIM-7 Sparrows and a centreline pylon. Optional fuselage pylons could also be fitted.

Initially, it was equipped with the Hughes AN/APG-63 radar, which did not require a second crewmember to operate, and a Head-Up Display to replace the traditional gunsight. Another innovation was Hands On Throttle And Stick or 'HOTAS', a system where the aircraft's most important controls are integrated into the pilot's joystick and throttle.

The first prototype F-15A made its first flight on July 27, 1972, and the first two-seater F-15B followed on October 18, 1973. A new single seater all-weather air-superiority version, the F-15C, was introduced in 1979. Externally it differed little from the original F-15A but there

▼ MCDONNELL DOUGLAS F-15A BAZ

McDonnell Douglas F-15A *Baz*, 646, 133 Squadron, Israel Defense Force – Air Force, Tel Nof Airbase, Israel, 1984.
The Israeli Air Force experimented with some different colour schemes; today the fleet continue to operate in the original colour schemes.

US JET FIGHTERS IN FOREIGN SERVICE — McDonnell Douglas/Boeing F-15 Eagle

▼ MCDONNELL DOUGLAS F-15DJ EAGLE

McDonnell Douglas F-15DJ Eagle, 12-8076, Tactical Fighter Training Group, Japan Air Self-Defense Force, Komatsu Air Base, Japan, 2019.
The TFTG uses a fleet of F-15s painted in special colour schemes, for aggressor training.

were significant changes internally. The undercarriage was strengthened to cope with a greater maximum weight – up to 68,000lb – and the F-100 engines were improved. The airframe itself was also strengthened, allowing the pilot to pull 9 g, compared to the previous limit of 7.5 g.

The AN/APG-63 benefitted from greatly increased processing power and additional internal fuel tank capacity was provided. The aircraft could be fitted with fixed external conformal tanks for an even greater fuel load if required. There was also a new ejection seat known as the Advanced Concept Ejection Seat or 'ACES II' which improved pilot survivability in the event of an urgent departure being required. Its two-seater counterpart was the F-15D.

The USAF launched the Enhanced Tactical Fighter programme in March 1981 with the goal of finding a direct replacement for the F-111. Options included the Panavia Tornado, General Dynamics' F-16XL and a new two-seater F-15 – the F-15E. This was declared the winner on February 24, 1984, mainly because it embodied a significantly lower development cost.

ISRAEL

Pilots from the IDF/AF evaluated the F-15B prototype, known at the time as the TF-15A, in 1974 and found it to be an exceptional aircraft. Israel placed an order for 25 F-15As the following year. Four were delivered in December 1976 under the Peace Fox project, followed by 19 more plus a pair of F-15Bs under Peace Fox II. Unsurprisingly the Israelis called the F-15A/B 'Baz', meaning Eagle.

The Baz would acquit itself well during its first taste of combat – a mission to provide top cover for other IDF/AF aircraft attacking terrorist bases in southern Lebanon on June 27, 1979. The F-15s and a number of IAI Kfirs were intercepted by Syrian MiG-21s but during the ensuing battle five of the MiGs were shot down for no Israeli losses. Another five MiGs were shot down for no losses

▼ MCDONNELL DOUGLAS/MITSUBISHI F-15J

McDonnell Douglas/Mitsubishi F-15J, 52-8850, 306th Tactical Fighter Squadron, Japanese Air Self-Defense Force, Komatsu Air Base, Ishikawa Prefecture, Japan, 2004.
Several F-15s were painted with special colour schemes to celebrate the 50th anniversary of the JASDF.

by IDF/AF F-15s on September 24, 1979, and the Baz claimed yet another MiG-21 destroyed on June 27, 1980.

Syria also operated the MiG-25 Foxbat and this type had so far proven impossible to intercept. On February 13, 1981, the IDF/AF laid a trap. A lone RF-4E was flown over Lebanon with Baz top cover. A MiG-25P duly scrambled for the intercept and was shot down by a Baz using a Sparrow missile – the first Foxbat ever to be lost in combat.

Less than four months later, on June 7, 1981, Israeli F-15s flew top cover for F-16s when they attacked and destroyed Iraq's nuclear reactor at Osirak near Baghdad, and another MiG-25 was downed by an F-15 while trying to intercept another Israeli RF-4E on July 29, 1981. Two Syrian MiG-23s were then shot down by Israeli F-15s in May 1982. All this, however, was just a prelude to the events of June 1982 – when Israeli F-15s, F-16s and Kfirs shot down 92 Syrian jet fighters during Operation Peace for Galilee.

Eighteen F-15Cs and eight F-15Ds were delivered under Peace Fox III, followed by five more F-15Ds – actually F-15E airframes – under Peace Fox IV once those types became available. Israel named these 'Akef' or Buzzard. After Desert Storm in 1991, 17 more F-15As were also delivered to Israel. The older F-15As and Bs were then upgraded to F-15C/D standard.

It was announced on January 27, 1994, that Israel would buy 21 examples of its own bespoke F-15E variant – the F-15I. This two-seater, known in service as the Ra'am (meaning Thunder), would include electronic components made in Israel to Israeli specifications and some would carry Sharpshooter and LANTIRN pods for night-fighting capability. Deliveries commenced in 1998 and were completed in September 1999. Today Israel operates 84 F-15A/B/C/D/I aircraft.

JAPAN
Officers from the JASDF evaluated the

US JET FIGHTERS IN FOREIGN SERVICE — McDonnell Douglas/Boeing F-15 Eagle

F-15A/B at Edwards EFB from June to July 1975 and before the end of the year it was announced that the F-15 had been chosen to replace Japan's fleet of F-104J Starfighters. Not only that, Japan acquired a licence to manufacture that F-15 with Mitsubishi being the prime contractor. The single seater, largely equivalent to the F-15C with certain sensitive US electronics replaced with Japanese equipment, would be designated F-15J while the two-seater was the F-15DJ.

The first two F-15Js were constructed by McDonnell Douglas in the US and first flew on June 4, 1980. The next eight were put together by Mitsubishi from MD-made kits.

When deliveries finally ceased in December 10, 1999, Japan had received 163 F-15Js and 36 F-15DJs made by Mitsubishi, plus the two MD-built F-15Js and 14 F-15DJs for a combined total of 215 Eagles. Over the years, eight Japanese F-15s have been lost in accidents.

SAUDI ARABIA

An order for 47 F-15Cs and 15 F-15Ds was placed by Saudi Arabia to replace its BAC Lightnings in 1980. The US agreed under project Peace Sun but on the condition that the Saudis could only have 60 F-15s in their country at any one time.

The first examples were delivered in August 1981. Three years later, Saudi F-15s fought off Iranian F-4Es that were threatening oil fields – downing two of them with Sparrow missiles. An order for nine new F-15Cs and three F-15Ds was placed in 1989 and deliveries began in mid-1991.

In the wake of the Gulf War, the Saudis sought a replacement for their Panavia Tornado strike fighters. And the F-15E's impressive performance during the conflict persuaded them that this was the right choice. President George Bush signed off on the purchase in late 1992 as project Peace Sun IX and McDonnell Douglas received a $122 million contract for 72 F-15XPs – the XP standing for 'export' – which was effectively a downgraded variant of the F-15E. By May 1993, the designation had been changed to F-15S, with the S standing for Saudi Arabia.

The first RSAF F-15S made its maiden flight on June 19, 1995, before being handed over to the Saudis that September. The last F-15S was delivered in 1999.

Just over a decade later, in 2010, Saudi Arabia ordered 84 examples of the F-15SA – an upgraded F-15S featuring a digital electronic warfare suite, fly-by-wire control system, AESA radar, infrared search-and-track system, advanced cockpit displays and Joint Helmet Mounted Cueing Systems. The last of these was delivered in December 2020.

▼ BOEING F-15QA *ABABIL*

Boeing F-15QA *Ababil*, QA-503, Ababil Squadron, Qatar Emiri Air Force, Qatar, 2022.
The F-15QA is a specific variant designed for Qatar and it uses this unique two-tone grey colour scheme; this aircraft carries AGM-84 Harpoon anti-ship missiles.

▼ BOEING F-15K SLAM EAGLE

Boeing F-15K Slam Eagle, 08-047, Republic of Korea Air Force, 11th Fighter Wing, Daegu Air Base, South Korea, 2021.
This aircraft carries the TAURUS KEPD 350 – a cruise missile designed to strike high-value fixed targets such as command and control facilities, hardened bunkers and airfields.

▼ MCDONNELL DOUGLAS F-15C EAGLE

McDonnell Douglas F-15C Eagle, 502, 13 Squadron, Royal Saudi Air Force, Dhahran Air base, Saudi Arabia, 1991.
Saudi F-15s participated in Operation Desert Storm in 1991, scoring two kills in air-to-air combat.

US JET FIGHTERS IN FOREIGN SERVICE — McDonnell Douglas/Boeing F-15 Eagle

▼ BOEING F-15SA EAGLE

Boeing F-15SA Eagle, 5513, 55 Squadron, Royal Saudi Air Force, King Khalid Air Base, Khamis Mushait, Saudi Arabia, 2014.
This F-15SA carries an impressive weapons load including, among others, the SLAM-ER cruise missile.

▼ BOEING F-15SG EAGLE

Boeing F-15SG Eagle, AF05-0007, Republic of Singapore Air Force Detachment – Peace Carvin V, 428th Fighter Squadron (USAF), Mountain Home Air Force Base, Idaho, USA, 2014.
The RSAF has F-15s operating from a USAF air base for training purposes; this aircraft's tail art commemorates the fifth anniversary of the Peace Carvin V partnership.

▼ MCDONNELL DOUGLAS/MITSHUBISHI F-15J EAGLE

McDonnell Douglas/MItshubishi F-15J Eagle, 62-8876, 305th Tactical Fighter Squadron, Japan Air Self-Defense Force, Nyutabaru Air Base, Japan, 2013.
A special paint scheme to commemorate the 40th anniversary of 305th Squadron.

Today the Royal Saudi Air Force operates 211 F-15C/S/SA types plus 21 F-15D conversion trainers. This makes Saudi Arabia the most numerically significant Eagle user outside the US – having now overtaken Japan.

SOUTH KOREA

A fighter acquisition programme for the Republic of Korea Air Force was launched in 1999 with the goal of replacing its aging F-4s. Four F-X contenders were selected – Boeing's F-15E, the Dassault Rafale, the Sukhoi Su-35 and the Eurofighter. The choice had been whittled down to the Rafale and F-15E by October 2000 – with the F-15 variant now being dubbed F-15K for Korea.

Ultimately the F-15K won and an order for 40 examples was placed, with the type being dubbed 'Slam Eagle'. Under the terms of the agreement, South Korean companies were responsible for producing 40% of the components including the fuselages, wings, flight control actuator, electronic jammer and radar warning receiver. Final assembly took place at Boeing's facility in St Louis. The first two were delivered in October 2005.

A second batch of 21 F-15Ks were ordered in April 2008, this time powered by Pratt & Whitney F100-PW-229 (EEP) engines made under licence in Korea by Samsung Techwin. The Republic of Korea Air Force currently operates 59 F-15K strike fighters, one having been lost during a night time training mission on June 7, 2006, and another having crashed into a mountain on April 5, 2018.

SINGAPORE

The government of Singapore began to seek a replacement for the Republic of Singapore Air Force's retired A-4SU attack aircraft in 2003. The contenders were narrowed down to the F-15E, Rafale and Eurofighter Typhoon. First the Typhoon was knocked out, then the Rafale, leaving the F-15 as the winner. A $1.6bn contract for 12 F-15Es was signed in December 2005. The first of these aircraft, initially dubbed F-15T, then F-15SG, was delivered in May 2009. Meanwhile, Singapore had ordered an additional eight in October 2007, with these being delivered between early 2016 and July 2017, then another four – for a total of 24. Then eight more were ordered in 2010 and another eight in 2014 – bringing the fleet up to 40 F-15SGs by 2018. As of October 2022, all 40 remain in service.

QATAR

The sale of up to 72 F-15Es to Qatar was approved by the US State Department in November 2016 and the following June Qatar agreed to buy 36 examples, to be known as F-15QA Ababil (Qatar Advanced; Ababil is a reference to a protective flock of birds mentioned in the Koran) in Qatar Emiri Air Force service, for $12bn. The first F-15QA made its flight debut on April 13, 2020, and the first four were delivered at the end of October 2021.

US JET FIGHTERS IN FOREIGN SERVICE

US JET FIGHTERS IN FOREIGN SERVICE — General Dynamics/Lockheed Martin F-16 Fighting Falcon

GENERAL DYNAMICS/ LOCKHEED MARTIN F-16 FIGHTING FALCON

The highly capable and thoroughly modern multirole F-16, commonly known as the Viper despite its 'official' name, has been the greatest US jet fighter success in recent times with hundreds of examples still in service today with air forces around the globe.

1979–PRESENT

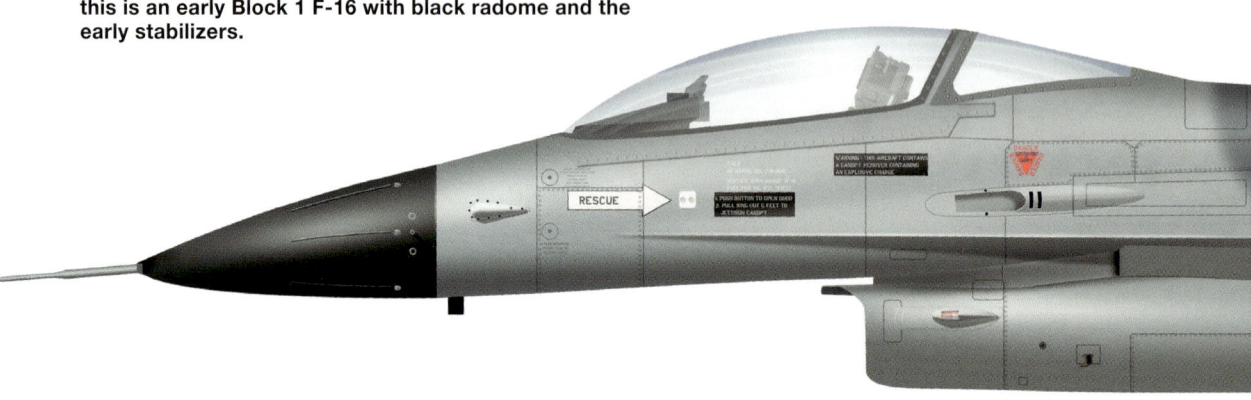

▼ GENERAL DYNAMICS/SABCA F-16A FIGHTING FALCON

General Dynamics/SABCA F-16A Fighting Falcon, FA-01, 349 Squadron, 10th Wing, *Belgische Luchtmacht – Force Aérienne Belge* (Belgian Air Force), Kleine Brogel Air Base, Belgium, 1980.
The 349th was the first non-US operational F-16 squadron; this is an early Block 1 F-16 with black radome and the early stabilizers.

combat experience during the 1960s changed the way the American military thought about jet fighters. Where previously it had been believed that a fighter had to be as fast as possible and armed only with missiles, there came a realisation that a lightweight and manoeuvrable platform could be equally successful, if not more so, and cheaper to boot.

The Air Force Prototype Study Group was formed in May 1971 and one of its proposals was the Lightweight Fighter (LWF). A request for proposals was drawn up calling for a 20,000lb day fighter optimised for combat at Mach 0.6-1.6 at altitudes of between 30,000ft and 40,000ft. It had to have excellent manoeuvrability, range and acceleration and ideally the cost of a single unit needed to be £3 million. This was issued on January 6, 1972.

Five companies offered designs and these were quickly narrowed down to Northrop's twin-engine P-600 and General Dynamics' single engine Model 401. Each company received a contract and funding to build two prototypes of its design, the designs being designated YF-17 and YF-16 respectively. The first YF-16 was rolled out on December 13, 1973, and its first official flight was on February 2, 1974 – although it had previously

▼ LOCKHEED MARTIN F-16C FIGHTING FALCON

Lockheed Martin F-16C Fighting Falcon, 210, 2nd Tactical Fighter Squadron, Royal Bahraini Air Force, Isa Air Base, Bahrain, 2016.
Bahraini F-16s operated alongside F-5s during Operation Desert Storm; new F-16Vs are to be acquired and the current F-16 fleet will be upgraded to F-16V standard.

US JET FIGHTERS IN FOREIGN SERVICE **095**

US JET FIGHTERS IN FOREIGN SERVICE — General Dynamics/Lockheed Martin F-16 Fighting Falcon

▼ LOCKHEED MARTIN F-16C FIGHTING FALCON

Lockheed Martin F-16C Fighting Falcon, 853, *Grupo nº 3, 1ª Brigada Aerea, Fuerza Aérea de Chile* (3rd Group, 1st Air Brigade, Chilean Air Force), Los Condores Air Base, Iquique, Chile, 2008.
Chile operates a mixed F-16 fleet of new F-16C/Ds and F-16A/Bs acquired from the Netherlands.

▼ GENERAL DYNAMICS/FOKKER F-16AM FIGHTING FALCON

General Dynamics/ Fokker F-16AM Fighting Falcon, E-191, Eskadrille 727, *Flyvevåbnet* (727 Squadron, Royal Danish Air Force), Skrydstrup Air Base, Denmark, 2019.
This colour scheme celebrates 800 years of the Danish Flag (*Dannebrog*).

▼ LOCKHEED MARTIN/TAI F-16C FIGHTING FALCON

Lockheed Martin/TAI F-16C Fighting Falcon, 9953, 272nd Tactical Fighter Wing, Egyptian Air Force, Gianaklis Air Base, Egypt, 2021.
Egypt is a major operator of the F-16, fielding several variants from the F-16A Block 15 to the F-16C Block 52; Egyptian F-16s have identification orange painted areas. Egypt's Block 40 F-16Cs can carry the AGM-84 Harpoon anti-ship missile.

become airborne for six minutes following a high-speed taxi test on January 20, 1974. The second prototype first flew on May 9, 1974.

The LWF competition soon became an acquisition programme with the advanced but inexpensive fighters attracting attention from other NATO air forces. The YF-16 proved to be more manoeuvrable than the YF-17 and had the same engine as the F-15, ensuring parts commonality and lowering costs. It was announced as the competition winner on January 13, 1975.

The first 'development' F-16A was rolled out on October 20, 1976, and made its debut flight on December 8. The first development F-16B two-seater began flight testing on August 8, 1977, and the first full production model F-16A took its first flight on August 7, 1978. The latter was delivered to the USAF on January 6, 1979, and the type received its official name 'Fighting Falcon' on July 21, 1980.

The single seat F-16C and its two-seat counterpart, the F-16D, were introduced in 1984 as Block 25 – with the first 'C' making its flying debut on June 19, 1984, and the type entering full production that December. These were powered by the Pratt & Whitney F100-PW-220E and featured improved avionics and radar, specifically the AN/APG-68, which allowed the use of beyond-visual-range AIM-7 and AIM-120 missiles. They also had improved ground-attack capability, being able to use the AIM-65D Maverick.

The next major update of the design came in 1986 with the F-16C/D Block 30/32. The biggest change was the switch to a new powerplant – the General Electric F110-GE-100 – for Block 30, with the Block 32 aircraft keeping the F100-PW-220E. The Block 30/32 aircraft could carry AGM-88A High-speed Anti-Radiation Missiles (HARMs) as well as the AGM-45 Shrike Anti-Radiation Missile (ARM). Avionics and decoy measures were also improved with twice the previous number of chaff/flare dispensers. The first example flew on June 12, 1986.

Block 40/42 was introduced in 1988. These have digital flight controls, replacing the old analogue system of the Block 25s, 30s and 32s. They also have an advanced holographic Head Up Display linked to a Martin Marietta LANTIRN (Low Altitude Navigation and Targeting Infra-Red, Night) targeting pod. Fitting this pod meant that the F-16's undercarriage legs actually had to be extended slightly for better ground clearance and the landing gear needed bigger wheels and tyres. The knock-on effect of this was a need for 'bulged' landing gear doors. This in turn meant that the aircraft's landing lights, which had been on the gear doors, had to be moved to the nose gear door.

An avionics upgrade means that Block 40/42 aircraft now have automatic terrain following and a new GPS navigation system plus new decoy launchers. The airframe itself was

US JET FIGHTERS IN FOREIGN SERVICE **097**

US JET FIGHTERS IN FOREIGN SERVICE — General Dynamics/Lockheed Martin F-16 Fighting Falcon

strengthened too, in order to cope with the additional weight of equipment. Block 40/42 F-16Cs and Ds can now use the Paveway family of guided weapons including the GBU-10, GBU-12 and GBU-24 Paveway laser-guided bombs and the GBU-15 glide bomb.

Pilots flying Block 40/42 F-16s are able to use night vision goggles and there is a data-link system which allows Forward Air Controllers to upload new data directly to the aircraft's weapons system computer which then puts it onto the pilot's HUD. Block 40/42 aircraft were also part of the Have Glass programme, intended to reduce their radar signature. This features a gold-tinted 'indium tin oxide' cockpit canopy and radar absorbent material including paint, making the aircraft up to 15% more difficult to detect.

The Block 50/52 is designed to complement the Block 40/42. This version features the Honeywell H-423 Ring Laser Gyro Inertial Navigation System (RLG INS), enabling quicker in-flight alignment, an enhanced GPS receiver, AN/ALR-56M advanced radar warning receivers and a Tracor AN/ALE-47 countermeasure system. While Block 50 machines are powered by the General Electric F110-GE-129, the Block 52s have the Pratt & Whitney F100-PW-229. The first Block 50/52 aircraft flew in October 1991.

The F-16V, unveiled in 2012, is the latest variant to date. It includes an AN/APG-83 active electronically scanned away (AESA) radar, upgraded mission computer and other cockpit improvements.

BELGIUM/DENMARK/NETHERLANDS/NORWAY

These four European NATO members formed a consortium on July 21, 1975, to work together with the US to build F-16s under licence. The original Belgian order was for 96 F-16As and 20 F-16Bs; Denmark wanted 46 As and 12 Bs; the Netherlands decided it needed 80 As and 22 Bs and Norway ordered 60 As and 12 Bs.

There were two production lines – one run by the primary Belgian contractor, Societe Anonyme Belge de Constructions Aeronautiques (SABCA), which would assemble the Belgian and Danish fighters, and one run by Fokker in the Netherlands with put together the Dutch and Norwegian fighters. Belgian contractor Fabrique National produced the F100 engines for all of the European-made F-16s.

The SABCA line got started in February 1978, followed by Fokker two months later. The first flight of a Belgian-built F-16, an F-16B, was on December 11, 1978, and the Belgian Air Force took delivery of its first F-16 on January 29, 1979, and its last from the initial batch in 1985. A follow-on order for 40 F-16As and four Bs was

LOCKHEED MARTIN F-16C FIGHTING FALCON

Lockheed Martin F-16C Fighting Falcon, 530, *337 Mira, Polemikí Aeroporía* (337 Squadron, Hellenic Air Force), Larisa Air Base, Greece, 2007.
Painted in the Aegean Ghost colour scheme, this Greek F-16 carries both the AIM-120 AMRAAM and IRIS-T air-to-air missiles; it has conformal fuel tanks installed.

LOCKHEED MARTIN F-16IQ (D) FIGHTING FALCON

Lockheed Martin F-16IQ (D) Fighting Falcon, 1620, 9th Fighter Squadron, Iraqi Air Force, Balad Air Base, Iraq, 2018.
Iraqi F-16s were acquired in the reconstruction phase of the air force, following the 2003 conflict; there are several reports casting doubts about the serviceability of the fleet.

▼ GENERAL DYNAMICS F-16A FIGHTING FALCON

General Dynamics F-16A Fighting Falcon, TS-1612, *Elang Biru* (Blue Eagle) demonstration team, *Skadron Udara 3, Tentara Nasional Indonesia Angkatan Udara* (3 Squadron, Indonesian Air Force), Iswahyudi Air Base, Java, Indonesia, 1995.
The *Elang Biru* was formed with six F-16s and performed from 1995 until 2000.

US JET FIGHTERS IN FOREIGN SERVICE
General Dynamics/Lockheed Martin F-16 Fighting Falcon

▼ **LOCKHEED MARTIN F-16I *SUFA***

Lockheed Martin F-16I *Sufa*, 470, 253 Squadron, Israeli Air Force, Ramon Air Base, Israel, 2020.
The F-16I is a variant designed for the specifications of the Israeli Air Force; this aircraft carries the AIM-120 AMRAAM and Rafael Python 5 air-to-air missiles, alongside Rafael Popeye air-to-ground missile.

placed in February 1983 with deliveries taking place from 1987 to 1991.

Denmark got its first F-16 from SABCA on January 28, 1980, and its last from the initial order in 1983. The Danes also placed a follow-on order, in August 1984, for eight F-16As and four F-16Bs, but these were built by Fokker rather than SABCA. Three American-made ex-Air National Guard F-16A attrition replacements were received in July 1994, followed by three more ex-ANG F-16As and one F-16B in early 1997.

The first Fokker-built F-16 made its flight debut on May 3, 1979, and the Royal Netherlands Air Force received its first aircraft in June of that year. Delivery of the initial batch was concluded in 1984 but a follow-on order had been approved in December 1983 for another 97 F-16A and 14 Bs. These were delivered from 1984-1992.

Norway's first F-16, off the Fokker line, made its debut flight on December 12, 1979, and deliveries took place from January 1980 to June 1984. There would be no follow-on for Norway however, just

▼ GENERAL DYNAMICS F-16A FIGHTING FALCON

General Dynamics F-16A Fighting Falcon, MM7244, *23° Gruppo, 5° Stormo, Aeronautica Militare* (23rd Group, 5th Wing, Italian Air Force), Cervia Air Base, Italy, 2010.
When the time came for Italy to return its leased F-16s several commemorative schemes were painted, including this stunning viper design.

two F-16Bs sent directly from the US as attrition replacements in 1989.

Today Belgium has just 40 upgraded F-16AMs, Denmark has 33 F-16AMs and 10 F-16BMs, the Netherlands has 40 F-16AMs and Norway retired its F-16 fleet entirely in 2021 – selling some to Romania and others to private US company Draken International.

ISRAEL

The US government initially resisted Israeli attempts to order the F-16 but relented in August 1978 when permission was given for the sale of 75 F-16A/Bs under the Peace Marble programme. The first four arrived in Israel during July 1980 and less than a year later they were already in combat – shooting down two Syrian Mi-8 helicopters on April 28, 1981. Then on June 7, 1981, eight IDF/AF F-16s destroyed Iraq's Osirak nuclear reactor near Baghdad. Then, from 1983-1984, F-16s were used in strikes on Syrian missile sites in Lebanon. They also destroyed 44 Syrian aircraft – with one F-16 managing to rack up six kills.

Under Peace Marble II, Israel order 51 F-16Cs and 24 F-16Ds, the first being received in October 1987. Peace Marble III commenced in 1988 and saw the delivery of 30 more F-16Cs plus 30 more F-16Ds. Peace Marble IV then provided Israel with 50 surplus F-16A/Bs, all deliveries taking place in 1994. Finally, 102 additional brand new F-16Ds were bought under Peace Marble V, with deliveries from 2003 to 2009. Today the F-16 continues to provide the backbone of the IDF/AF with 175 single seaters and 49 two-seaters in its inventory.

SOUTH KOREA

The Republic of Korea Air Force was the first foreign operator of the F-16C/D variants, the South Korean government having ordered 30 Cs and six Ds in December 1981. Deliveries were made between 1986 and 1988, with a further four F-16Ds being ordered under the same programme, Peace Bridge, in 1988. The much more ambitious Peace Bridge II then saw a huge order placed for 80 F-16Cs and 40 F-16Ds. This also encompassed offset measures where the first 12 would be made by Lockheed, the next 36 assembled in Korea from kits and the remaining 72 built entirely in Korea by Samsung Aerospace. The US-made examples arrived on December 2, 1994, and the first five F-16s made from Lockheed kits were accepted on November 9, 1995.

Korean Aerospace Industries then made 20 more Block 52 F-16C/Ds under the Peace Bridge III programme, starting in July 2000, with deliveries from 2003-2004. At the time of writing the RoKAF was able to field up to 167 F-16C/D/V variants.

EGYPT

Under the Peace Vector programme, the Egyptian government ordered 34 F-16As and eight F-16Bs on June 25, 1980. The first was delivered to Egypt in March 1982. Peace Vector II saw 34 more F-16Cs and six Ds built for Egypt with deliveries beginning in October 1986. Then Peace Vector III of June 1990 resulted in 35 more F-16Cs and 12 F-16Ds being bought, arriving from October 1991.

Peace Vector IV followed but this time the aircraft were produced in Turkey by TAI – 34 F-16Cs and 12 F-16Ds – with deliveries starting in early 1994. Peace Vector V was for 21 new F-16Cs, delivered from 1999 to 2000, Peace Vector V was for 12 Cs and 12 Ds, delivered from 2001-2002, and Peace Vector VII saw another 16 Cs and four Ds delivered from 2012-2015. Today Egypt has 218 F-16A/B/C/Ds.

US JET FIGHTERS IN FOREIGN SERVICE

General Dynamics/Lockheed Martin F-16 Fighting Falcon

▼ MITSUBISHI F-2A

Mitsubishi F-2A, 03-8559, 3rd Squadron, Japanese Air Self Defense Force, Hyakuri Air Base, Japan, 2017.
The Mitsubishi F-2 is a fighter derived from the F-16, externally distinguishable by a bigger wing and stabilizers and a three-piece canopy; it features several other differences in the avionics and weapons characteristics; this aircraft carries indigenous weapons in the form of AAM-3 and AAM-4 air-to-air missiles and the ASM-2 anti-ship missile.

▼ GENERAL DYNAMICS F-16A FIGHTING FALCON

General Dynamics F-16A Fighting Falcon, 717, 14th Squadron, Pakistan Air Force, Minhas Air Base, Pakistan, 1988.
Pakistani F-16s saw plenty of action countering border violations by Soviet aircraft operating in neighbouring Afghanistan; this aircraft carries two variants of the AIM-9 air-to-air missile.

▼ GENERAL DYNAMICS/FOKKER F-16BM FIGHTING FALCON

General Dynamics/Fokker F-16BM Fighting Falcon, J-368, 311 Squadron, *Koninklijke Luchtmacht* (Royal Netherlands Air Force), Volkel Air Base, The Netherlands, 2003.
This F-16B carries the MARS reconnaissance pod on the centreline station.

PAKISTAN

A letter of offer and acceptance on 28 F-16As and 12 F-16Bs was signed in December 1981 as the Peace Gate programme. All 40 aircraft had been delivered by mid-1986 and between May 1986 and November 1988 Pakistani F-16s shot down at least eight intruders from Afghanistan. These included four Su-22s, two MiG-23s, one A-26 and one Su-25. During these encounters one F-16 was lost after being accidentally hit by the Sidewinder of another F-16.

Another 71 aircraft were ordered in December 1988 but this order was suspended when the US government placed an arms embargo on Pakistan in October 1990. While the Pakistanis argued their case, work on building the aircraft continued with 28 being produced and put into storage. The remaining 43 would never be constructed. Eventually, the 28 would be brought into service as adversaries/aggressors for the US Navy and USAF before 26 survivors were eventually modernised and sent to Pakistan in 2006 after the embargo was lifted.

Under the Peace Drive programme of 2005, Pakistan has also been able to acquire 18 new aircraft – 12 F-16Cs and six F-16Ds. In 2014, another nine F-16As and four F-16Bs were bought from Jordan. Today Pakistan has a total of 75 operational F-16s of all types.

VENEZUELA

The Venezuelan government signed a deal to buy 18 F-16As and six F-16Bs to replace its existing fleet of Mirage IIIs and 5s in May 1982. The Peace Delta programme was approved in 1983 and the first aircraft was delivered that September. Venezuelan F-16 pilots sided with the government during an attempted coup in November 1992 and achieved at least one kill. Three of these aircraft have since been crashed and an arms embargo imposed on Venezuela by the US in 2006 is thought to have grounded the remainder. As of 2022, the Bolivarian National Air Force of Venezuela is believed to have 16 F-16As and four F-16Bs remaining.

TURKEY

Having been a significant user of the F-104 and F-5, Turkey was keen to place a large order for the F-16. It announced plans to buy 132 F-16Cs and 28 F-16Ds in September 1983. Under the terms of the Peace Onyx programme, Turkey would receive its first eight examples from the US but the other 152 would be assembled from kits locally at Mürted by TUSAS Aerospace Industries (TAI). TAI-assembled aircraft would not go straight to the Turkish Air Force however, they would be delivered to the US first before being sent on.

The first US-built F-16s were received in July 1987 and the first Turkish-built F-16C made its flight debut on October 20, 1987. Up to the 43rd aircraft, Turkish F-16s were made to Block 30 standard. From the 44th they were Block 40.

US JET FIGHTERS IN FOREIGN SERVICE

General Dynamics/Lockheed Martin F-16 Fighting Falcon

▼ LOCKHEED MARTIN F-16C FIGHTING FALCON

Lockheed Martin F-16C Fighting Falcon, 4056, Tiger Demonstration Team, *6 Eskadra Lotnictwa Taktycznego, Siły Powietrzne* (6th Tactical Aviation Squadron, Polish Air Force), RAF Fairford, 2016.
The demonstration aircraft is seen here as it performed at the 2016 Royal International Air Tattoo.

▼ LOCKHEED MARTIN F-16AM FIGHTING FALCON

Lockheed Martin F-16AM Fighting Falcon, 15116, *Esquadra 301, Força Aérea Portuguesa* (301 Squadron, Portuguese Air Force), Air Base 11, Beja, Portugal, 2021.
For the 2021 NATO Tiger Meet the PoAF had this stunning colour scheme applied to one of its F-16s.

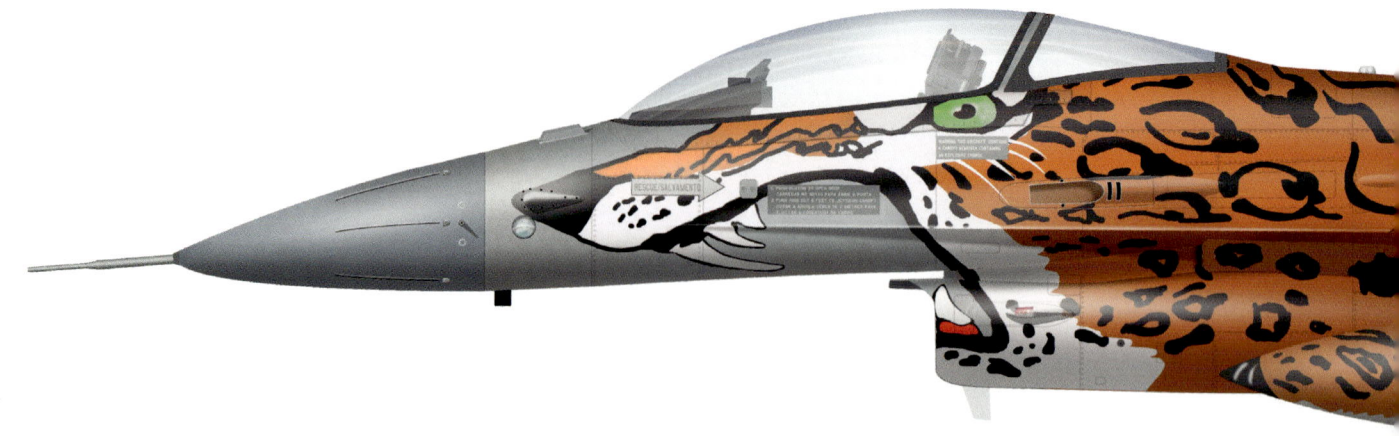

In addition to building 152 Block 30/40 F-16C/Ds for Turkey, TAI also made 34 Block 40 F-16Cs and 12 Block 40 Ds for Egypt. A new order for 68 F-16Cs and 12 F-16Ds, Peace Onyx II, was placed in March 1992 with the first examples being delivered in July 1996. Peace Onyx III, for another 40 F-16C/Ds, followed with deliveries between 1998 and 1999. Peace Onyx IV saw another 14 F-16Cs and 16 F-16Ds delivered from 2011 to 2012.

A Greek Mirage F1 crashed during a dogfight with a Turkish F-16 on June 18, 1992, and Turkish F-16s were flown in support of UN operations over Bosnia and Herzegovina and Kosovo in 1993. A Turkish F-16 was then lost on February 8, 1995, when it crashed into the Aegean Sea after being intercepted by Greek F1s. The following year, on October 8, a Greek Mirage 2000 shot down a Turkish F-16D with an R.550 Magic II missile. The co-pilot ejected and was rescued by Greek forces but the pilot was killed. A Turkish F-16 collided with a Greek F-16 on May 23, 2006, and while the Turkish pilot ejected

safely the Greek pilot was killed. And in March 2014 a Turkish F-16 shot down a Syrian MiG-23 for violating Turkish airspace.

Today Turkey operates 158 F-16Cs and 87 F-16Ds – a total of 245.

SINGAPORE

Under the Peace Carvin programme, initiated in January 1985, Singapore placed an order for four F-16As and the same number of F-16Bs, with an option for 12 more. Deliveries began on February 20, 1988, though the aircraft did not actually arrive in Singapore till January 1990. Peace Carvin II saw eight F-16Cs and 10 F-16Ds ordered on July 9, 1994, with deliveries starting during the spring of 1998. At the same time, a commercial deal was signed under lend-lease terms for four more F-16Cs and eight more F-16Ds to replace the original F-16A/Bs.

Peace Carvin III and Peace Carvin IV then saw another 10 F-16Cs and 22 F-16Ds delivered from 2000 to 2004. At the time of writing, the Republic of

US JET FIGHTERS IN FOREIGN SERVICE **105**

US JET FIGHTERS IN FOREIGN SERVICE
General Dynamics/Lockheed Martin F-16 Fighting Falcon

▼ **GENERAL DYNAMICS F-16AM FIGHTING FALCON**

General Dynamics F-16AM Fighting Falcon, 1601, *Escadrila 53 Vânătoare, Forțele Aeriene Române* (53 Fighter Squadron, Romanian Air Force), 86th Air Base, Fetesti, Romania, 2016. Romanian F-16s use a unique three-tone grey colour scheme. The aircraft were acquired via Portugal, both from the PoAF's own inventory and former USAF aircraft that were upgraded in Portugal prior to delivery to the FAR.

▼ **LOCKHEED MARTIN F-16C FIGHTING FALCON**

Lockheed Martin F-16C Fighting Falcon, 1, Black Knights demonstration team, Republic of Singapore Air Force, Tengah Air Base, Singapore, 2016.
The Black Knights introduced the F-16 in 2000.

Singapore Air Force had 60 F-16C/Ds on strength.

THAILAND
The government of Thailand attempted to order eight F-16As and four F-16Bs in 1985 but the order was not approved until July 1987. The first F-16A was formally handed over in May 1988 under the Peace Naresuan programme. An additional six F-16As were then delivered under Peace Naresuan II from 1990 to 1991. Peace Naresuan III saw another 12 F-16As and six F-16Bs delivered from 1995 to 1996, followed by Peace Naresuan IV in 2000 which involved the purchase of 15 more F-16As and one more B. Finally, in 2004, Singapore donated three of its F-16As and four F-16Bs to Thailand in exchange for the opportunity to train at a Thai base for a certain number of days per year. Today Thailand continues to operate 37 F-16As and 14 F-16Bs.

GREECE
Turkey bought F-16s so Greece followed suit. An initial order for 34 F-16Cs and six F-16Ds was announced in November 1984 but it wasn't signed

till January 1987 due to the Greek government haggling over the price and the US expressing concerns about technology transfer to Warsaw Pact countries. The aircraft were delivered under the Peace Xenia programme from November 1988 to October 1989. Peace Xenia II was initiated in April 1993, with another 32 F-16Cs and eight more Ds on order. The first aircraft was formally accepted in May 1997. Then, in June 2000, Peace Xenia III saw another 34 new F-16Cs and 16 F-16Ds ordered with deliveries from 2002 to 2004. Finally, on December 13, 2005, Greece ordered yet another 20 F-16Cs and 10 F-16Ds under Peace Xenia IV, these being delivered from 2009 to 2010. Today Greece has 115 F-16Cs and 39 F-16Ds.

As mentioned in the Turkey section, Greek F-16s have clashed with Turkish forces on several occasions.

INDONESIA
Looking to replace some of its old MiG-21s, Indonesia signed up to purchase eight F-16As and four F-16Bs in August 1986 under the Peace Bima-Sena programme. Deliveries began on December 11, 1989, and were concluded the following year. Attempts were made to buy more during the 1990s but these came to nothing. Eventually however, under Peace Bima-Sena II, the Indonesians were able to buy 19 ex-USAF F-16Cs and five F-16Ds in 2014, deliveries being completed by the end of 2015. Today Indonesia still operates 33 F-16A/B/C/Ds.

BAHRAIN
A wide range of future fighter options were considered by Bahrain – including the F-15, F/A-18, Tornado, Mirage 2000 and various Russian types – before it settled on the F-16. A letter of agreement was signed for eight F-16Cs and four F-16Ds in March 1987 under Peace Crown. The first of these was formally received on March 22, 1989, and Bahrani F-16s joined Coalition forces during Desert Storm in 1991. A further agreement was signed in April 1998 for 10 F-16C Block 40s equipped with LANTIRN under Peace Crown II and these were delivered from 2000 to 2001. It was agreed in 2017 that Bahrain would pay for 20 of its existing F-16 Block 40s to be upgraded to F-16V Block 70 standard. As of 2021 Bahrain was planning to also buy 16 new-build Block 70s. The first Block 70s were due to be delivered in 2024 but it is not clear whether these will be the upgraded existing aircraft or new builds.

TAIWAN
A deal was struck to supply the Republic of China Air Force with F-16s under President George Bush in 1992 as Peace Phoenix. A total of 120 F-16A and 30 F-16B Block 20s were ordered and deliveries took place between April 1997 and 2001. The entire fleet was due to be upgraded to F-16V standard starting in 2016 – though how far this had progressed by the time of writing is unclear. It was reported in 2020 that a further 66 new-build F-16Vs had been ordered by Taiwan with deliveries expected to commence in 2026.

PORTUGAL
A letter of offer and acceptance for 20 new-build aircraft, 17 F-16As and three F-16Bs, was signed by Portugal in 1990, with the first four being delivered in July 1994 under the Peace Atlantis Program. Another 21 F-16As and four F-16Bs, all ex-USAF, were bought in 1999 under Peace Atlantis II. At the same time, 20 upgrade kits were also purchased – bringing 16 of the F-16As and all of the F-16Bs up to the same Block 15 OCU standard as the first

US JET FIGHTERS IN FOREIGN SERVICE
General Dynamics/Lockheed Martin F-16 Fighting Falcon

▼ LOCKHEED MARTIN F-16V FIGHTING FALCON

Lockheed Martin F-16V Fighting Falcon, 6613, 21st Tactical Fighter Group, 4th Tactical Fighter Wing, Republic of China Air Force, Chiayi Air Base, Taiwan, 2021.
The RoCAF declared the world's first F-16V unit operational in November 2021.

▼ GENERAL DYNAMICS F-16A FIGHTING FALCON

General Dynamics F-16A Fighting Falcon, 40317, 403 Squadron, 4th Wing, Royal Thai Air Force, Takhli Air Base, Thailand, 2015.
This RTAF F-16 carries special tail art to celebrate the 20th anniversary of 403 Squadron.

20 aircraft. A dozen Portuguese F-16s were sold to Romania in 2012, followed by another five in 2019. Today Portugal operates 27 F-16s, one having been lost in a crash in 2008.

JORDAN
A five-year lease on 12 F-16As and four F-16Bs, all modernised and refurbished ex-USAF examples, was signed by Jordan in July 1996 with an option to eventually purchase them. The first ones were delivered in December 1997 under the Peace Falcon programme. Congress gave approval on June 14, 2000, for a further 13 F-16As and four F-16Bs to be delivered under Peace Falcon II, starting in January 2003.

Jordan next bought 16 ex-Belgian Air Force F-16AM/BMs under Peace Falcon III in 2005, six ex-Dutch F-16BMs under Peace Falcon IV in 2009 and nine more ex-Belgian F-16AM/BMs under Peace Falcon V in 2011.

A Jordanian F-16 was allegedly shot down by ISIS on December 24, 2014, crashing in Syria. Its pilot, First Lieutenant Muath Safi Yousef al-Kasasbeh, was captured by ISIS militants and a video was released on February 3, 2015, showing him being burned alive. Today Jordan operates 44 F-16As and 15 F-16Bs.

ITALY
Delays to the Eurofighter Typhoon programme prompted the Italians to lease 30 refurbished ex-USAF F-16s as a stopgap in 2001 under the Peace Ceasar programme. These included 26 F-16A

fuel tanks. A total of 80 aircraft were delivered from 2003 to 2006 – 55 F-16Es and 25 F-16Fs. Some of these participated in strikes against Islamic State forces during February 2015.

OMAN

Under the Peace A'sama A'safiya (Clear Skies) I programme, the Sultanate of Oman purchased eight F-16C and four F-16D Block 50s. The Royal Air Force of Oman accepted the first of those on August 4, 2005. Deliveries continued into 2006. A new order for 10 F-16Cs and two F-16Ds was announced on August 3, 2010, and these were delivered in 2014 under Peace A'sama A'safiya II. A single F-16C from the original batch was lost in a fatal crash during a training exercise on September 22, 2013, leaving Oman with a total of 23 F-16s today.

CHILE

When it came to selecting a new fighter in 1997, Chile considered the SAAB Gripen, Mirage 2000 and F/A-18 before officially announcing on December 27, 2000, that the F-16C/D Block 50 had been chosen. An order was duly placed for six F-16Cs and four fully combat capable F-16Ds on February 1, 2002, to be delivered under the Peace Puma programme. The first examples began to arrive in June 2005 and in October that same year Chile bought 11 F-16AMs and seven F-16BMs from the Netherlands under the Peace Amstel I programme. These were delivered from 2006 to 2007. Finally, in May 2009 it was confirmed that Chile would buy 18 more Dutch F-16As. Deliveries were from 2010 to mid-2011 under Peace Amstel II. Today Chile has 35 F-16A/Cs and 11 F-16B/Ds.

POLAND

When Poland joined NATO in April 1999 its air force was still operating a fleet of MiG-21s and Western replacements were urgently needed. The Mirage 2000-9 and SAAB Gripen were considered but in the end the Poles chose the F-16 as their future fighter in December 2002. Under the Peace Sky programme, a letter of offer and acceptance was signed on March 15, 2003, for 36 F-16C and 12 F-16D Block 52s. Deliveries began in November 2006 and concluded in December 2008.

MOROCCO

Morocco ordered 24 F-16C/D fighters in 2007 and deliveries took place from July 2011 to August 2012. One of these was lost in 2015 while taking part in Saudi-led air operations in Yemen. A second order, for 25 new F-16C/D Block 72s, was approved in March 2019. Deliveries are due to commence in 2025.

ROMANIA

Romania joined NATO on March 29, 2004, and just under six years later, on March 24, 2010, the Romanian government made its first attempt to buy F-16s – 24 refurbished ex-USAF F-16C/D Block 25s. The deal fell through when the

ADFs, one F-16B ADF and three F-16B Block 5/10s. A further four aircraft, F-16A Block 10s, were also supplied to act as a source of spares. Deliveries took place from July 2003 to November 2004. With Eurofighter deliveries under way, Italy began to return its Falcons to the US in June 2010. The last one went home on May 23, 2012.

UNITED ARAB EMIRATES

The UAE embarked on a selection process for a new fighter in 1996 and soon narrowed the contenders down to the F-16 and Rafale. The F-16 was declared the winner on May 12, 1998, but there were disagreements about access to the technology and source code of the model in question – the new Block 60. Eventually, however, the deal was signed and the first UAEAF F-16 Block 60 flew on December 6, 2003. The variant was given the designation F-16E/F in recognition of significant upgrades made to its radar and avionics as well as the inclusion of conformal

US JET FIGHTERS IN FOREIGN SERVICE
General Dynamics/Lockheed Martin F-16 Fighting Falcon

▼ LOCKHEED MARTIN/TAI F-16C FIGHTING FALCON

Lockheed Martin/TAI F-16C Fighting Falcon, 91-0008, *192 Filo, Türk Hava Kuvvetleri* (192 Squadron, Turkish Air Force), 9th Main Jet Air Base, Balikesir, Turkey, 2015.
This aircraft carries a kill mark on its nose from an engagement with a Syrian MiG-23 in 2014, when it was flown by 182 Squadron.

▼ LOCKHEED MARTIN F-16E DESERT FALCON

Lockheed Martin F-16E Desert Falcon, 3060, 1st Squadron, United Arab Emirates Air Force, Al Dhafra Air Base, United Arab Emirates, 2009.
The F-16E Desert Falcon is a variant specific to the UAE; it features improved radar and avionics and can carry conformal fuel tanks.

▶ GENERAL DYNAMICS F-16A FIGHTING FALCON

General Dynamics F-16A Fighting Falcon, 1041, *Grupo Aereo de Caza 16, Aviación Militar Nacional Bolivariana* (Fighter Air Group 16, Venezuelan Air Force), El Libertador Air Base, Maracay, Venezuela, 2013.
This tail art was applied to celebrate the 30th anniversary of the Grupo Aereo de Caza 16 Dragones (Dragons); Venezuelan F-16s are painted in the South-East Asian colour scheme.

first scheduled payment was not made on time due to financial difficulties. The Romanians tried again in 2015, with a plan to purchase F-16s from Portugal. This time everything went as expected and Romania acquired nine F-16AMs and three F-16BMs, all of which were updated to the latest MLU standards. A second deal in 2019 saw Portugal sell another five F-16As to Romania, deliveries being completed in 2021. A plan to buy 32 ex-Norwegian F-16s for an estimated 454 million euros was approved in June 2022.

IRAQ
Concerned about perceived threats from Iran and Syria, the government of Iraq ordered 18 F-16C/D Block 50/52s in 2011 to replace its legacy MiG fighters. A second order for 18 was placed in 2012. The first of Iraqi F-16s, designated F-16IQ Block 52, were delivered on June 5, 2014, with deliveries being completed by May 2019 for a total of 36 F-16IQs – though two crashed during training.

SLOVAKIA
As a NATO member, Slovakia was able to sign a deal for the purchase of 14 F-16Vs in 2018. Deliveries are due to commence in 2024 – a year later than planned due to the COVID-19 pandemic and global microchip shortage.

BULGARIA
An order for eight F-16 Block 70s was placed by Bulgaria in 2019. They are due for delivery in 2025 and at the time of writing the NATO country was considering the purchase of a second batch of eight to make a full squadron of 16 aircraft.

US JET FIGHTERS IN FOREIGN SERVICE — McDonnell Douglas/Boeing F/A-18A/B/C/D

MCDONNELL DOUGLAS/BOEING F/A-18A/B/C/D

Doomed to play second fiddle to the outstanding F-16, the lightweight, cost-effective and perhaps most importantly twin-engined F/A-18 nevertheless managed a respectable export sales tally.

1982–PRESENT

▼ MCDONNELL DOUGLAS CF-188A HORNET

McDonnell Douglas CF-188A Hornet, 188761, 409 Squadron, Royal Canadian Air Force/ Aviation Royale Canadienne, Baden-Soellingen Air Base, West Germany, 1987.
RCAF Hornets were a major part of NATO aerial forces facing the Warsaw Pact in West Germany during the 1980s. RCAF Hornets had the fake canopy painted on the forward underside fuselage.

Having created the F-5, Northrop continued to develop its basic concept of a simple lightweight supersonic fighter into the 60s and early 70s. Extensive wind tunnel testing and refinement resulted in the P-530 Cobra – so-called because the substantial Leading Edge Root eXtensions (LERX) on its wings gave it a hooded appearance.

The design was showcased at the 1971 Paris Air Show and the following year it was entered for the USAF's new Lightweight Fighter (LWF) contest, which was intended to find a smaller aircraft to complement the F-15. The USAF subsequently picked the Northrop design alongside that of General Dynamics to develop as prototypes under the designations YF-17 and YF-16 respectively.

The first YF-17 flew on June 9, 1974, being joined by the second on August 21 of the same year. Each was powered by a pair of YJ101-GE-100 turbofans producing 14,400lb of thrust with afterburner. However, the YF-16 was chosen as the winner by a close margin, becoming the famous F-16.

Meanwhile, the Navy had been pushed to seek its own LWF which could complement the heavier F-14 as well as replacing A-7s and F-4s in the strike-fighter role. This resulted in what became the Navy ACF or 'NACF' competition, with McDonnell Douglas joining Northrop to navalise the YF-17.

The Navy announced on May 2, 1975, that the YF-17 was the winner and would be built as the F-18. McDonnell Douglas and Northrop signed an agreement that MD would be the lead constructor of the naval F-18 while Northrop would lead on the land version, the F-18L – with which they still hoped to attract many international customers.

Unfortunately for Northrop, the F-18L found no takers and the aircraft they had designed and developed was now sold to the US Navy with McDonnell Douglas as the manufacturer. Compared to the YF-17, the F-18 had bigger wings, more powerful General Electric F404 engines, increased fuel capacity, in-flight refuelling capability and high-lift devices for deck landings.

▼ MCDONNELL DOUGLAS CF-188A HORNET

McDonnell Douglas CF-188A Hornet, 188764, 439 Squadron, Royal Canadian Air Force/ Aviation Royale Canadienne, Albacete Air Base, Los Llanos, Spain, 1992.
This Hornet is properly painted for the 1992 NATO Tiger Meet.

US JET FIGHTERS IN FOREIGN SERVICE — McDonnell Douglas/Boeing F/A-18A/B/C/D

▼ MCDONNELL DOUGLAS F/A-18D HORNET

McDonnell Douglas F/A-18D Hornet, HN-466, *Hävittäjälentolaivue 31, Ilmavoimat* (31 Squadron, Finnish Air Force), Kuopio Air Base, Finland, 2018.
Finland carried out a second mid-life update on its Hornet fleet in 2012-2016; the aircraft, among other improvements, could now carry advanced weapons such as the JSOW and AIM-9X shown here.

▼ MCDONNELL DOUGLAS F/A-18C HORNET

McDonnell Douglas F/A-18C Hornet, 425, 9 Squadron, Kuwait Air Force, Ahmad al-Jaber Air Base, Kuwait, 1993.
Kuwaiti Hornets are painted in this unique three-tone grey colour scheme; they are now being replaced by F/A-18 Super Hornets and Eurofighter Typhoons.

The pilot sat on a Martin-Baker Mark 10 ejection seat and faced three multifunction displays linked to the aircraft's Hughes AN/APG-65 radar, which could be toggled between navigation, air combat and strike modes. The aircraft also featured an AN/ALQ-126B jamming system, AN/ALE-39 chaff and flare dispensers and an AN/ALR-50 radar warning system, plus the usual radio, IFF and navigation beacon.

Built-in armament was a single M61A1 Vulcan 20mm cannon in the top of the nose. Each wingtip had a single Sidewinder launch rail and there were two pylons for external stores beneath each wing. In addition, a single Sparrow-type missile could be recessed into the fuselage on either side and there was a centreline fuselage hardpoint too.

The first prototype flew on November 18, 1978 and the testing and evaluation programme which followed lasted until October 1982, by which time the type had been redesignated F/A-18 as an indicator of its multirole capability. The F/A-18B was the two-seat variant with a second cockpit behind the first and dual controls.

During 1987, McDonnell Douglas replaced the single-seat F/A-18A and two-seat F/A-18B on its production line with the single-seat F/A-18C and two-seat F/A-18D. Visually there was little change, except for the addition of a single strake to the rear of each LERX. These were retrofitted to all F/A-18As and Bs so it remained very difficult to tell the newer aircraft apart from their older siblings.

Internally, the F/A-18C differed from the F/A-18A in having an improved ejection seat and electronics upgrades.

The F/A-18D was essentially a two-seat F/A-18C – replicating the original F/A-18A and B format. The F/A-18C prototype – a modified F/A-18A – first flew on September 3, 1986.

CANADA

The Canadian government started looking for a single new fighter type that could replace the RCAF's CF-101 Voodoos, CF-116 Freedom Fighters and CF-104 Starfighters during March 1977. The choice was very quickly narrowed down to either the F-16 or the F/A-18.

It was announced on April 10, 1980, that the Hornet was the winner and an initial order was placed for 113 single-seaters plus 24 two-seaters – a total of 137. The deal included an option to buy up to 20 more and indeed an intention to purchase a further 11 single-seaters was indicated shortly thereafter.

Canadian F/A-18s are basically the same as those formerly used by the US Navy, except for the inclusion of an Instrument Landing System (ILS) instead of the naval F/A-18's Automatic Carrier Landing System (ACLS) and the fitment of a 600,000 candlepower spotlight on the port side fuselage to aid with the identification of other aircraft at night.

The single seaters were designated CF-188 while the two-seaters were CF-188Bs and the contract was altered to 98 single-seaters and 40 two-seaters for a total of 138. The option to buy the extra 11 was allowed to lapse in April 1985.

The first CF-188 flew on July 29, 1982, before being delivered on October 27. The last was delivered in September 1988. Twenty CF-188s were involved in Desert Storm in 1991 and all survived.

US JET FIGHTERS IN FOREIGN SERVICE — McDonnell Douglas/Boeing F/A-18A/B/C/D

▼ BOEING F/A-18D HORNET
Boeing F/A-18D Hornet, M45-02, *18 Skadron, Tentera Udara Diraja Malaysia* (18 Squadron, Royal Malaysian Air Force), Butterworth Air Base, Malaysia, 2017.
One of the most formidable weapons at the disposable of the Malaysian Hornet is the AGM-84 Harpoon anti-ship missile.

▼ MCDONNELL DOUGLAS EF-18M HORNET
McDonnell Douglas EF-18M Hornet, 15-01, *Ala 15, Ejército del Aire* (Wing 15, Spanish Air Force), Zaragoza Air Base, Spain, 2016.
This Spanish Hornet had a special painting for the 2016 NATO Tiger Meet; it carries IRIS-T air-to-air missiles.

Today Canada continues to fly 72 CF-188s and 31 CF-188Bs, plus 12 ex-RAAF F-18As and six F/A-18Bs purchased in 2018 and delivered by May 2021.

AUSTRALIA
When it came to replacing its Dassault Mirage IIIOs in 1975, Australia eventually faced much the same decision as Canada – the choice coming down to either the F-16 or the F/A-18. The F/A-18 was officially announced as the winner on October 20, 1981, and an initial order for 57 F/A-18As and 18 F/A-18Bs was placed. As part of the deal, it was arranged that 40% of the components for these aircraft would be made in Australia and assembly of all but the earliest examples would take place at the Government Aircraft Factory, later known as Aerospace Technologies of Australia (ASTA) at Avalon in Victoria.

The first two Australian Hornets, both F/A-18Bs, were made by McDonnell in the US and delivered on October 29, 1984. The first Australian assembled example made its flight debut on February 26, 1985, and was delivered on May 4 that year.

Like the Canadian CF-188s, the Australian Hornets were very similar to those of the US Navy and Marine Corps – except for the installation of ILS, a fatigue recorder and a high-frequency radio. The last one was delivered on May 16, 1990 and ASTA then switched to a

programme of upgrades for the aircraft. It would eventually form the nucleus of Boeing Australia.

Aussie Hornets were deployed against the Taliban in Afghanistan during 2001 and against ISIL in Iraq during 2015 as part of Operation Okra. The fleet was formally retired in November 2021.

SPAIN

The US government initially gave Spain the option to purchase 72 F-16s in May 1982 but in December it was announced that the Spanish had chosen to buy the F/A-18 instead – with the initial intention to buy 72 F/A-18As and 12 F/A-18Bs. This was then reduced to 60 F/A-18As and 12 F/A-18Bs in May 1983.

The first Spanish Hornet made its maiden flight on December 4, 1985, and the first F/A-18B was flown to Spain on July 10, 1986. In Spanish service, the aircraft are known as EF-18As and EF-18Bs – the E standing for Espana, 'Spain'. They were upgraded to C/D standard during the 1990s and in late 1995 the Spanish government bought 24 ex-US Navy surplus F/A-18A/Bs to supplement its fleet.

Nevertheless, by 2002 only six EF-18s had been lost in accidents – the best safety record of any Spanish Air Force fighter.

Over the years, EF-18s have served in Yugoslavia, Bosnia and Libya. Today Spain still operates 72 EF-18As and 12 EF-18BM conversion trainers.

US JET FIGHTERS IN FOREIGN SERVICE — McDonnell Douglas/Boeing F/A-18A/B/C/D

▼ BOEING F/A-18C HORNET

Boeing F/A-18C Hornet, J-5011, *Fliegerstaffel 11, Fliegergeschwader 13, Schweizer Luftwaffe/Forces Aériennes Suisses/Forze Aeree Svizzere* (11 Squadron, 13 Wing, Swiss Air Force), Meiringen Air Base, Switzerland, 2018.
One of the annual tasks for the Swiss Hornets is to enforce a no-fly zone during the Davos world economic forum; this aircraft carries the emergency radio frequency painted on its ventral tanks.

▼ McDONNELL DOUGLAS F/A-18A HORNET

McDonnell Douglas F/A-18A Hornet, A21-018, No. 75 Squadron, Royal Australian Air Force, RAAF Base Tindal, Northern Territory, Australia, 2021.
No. 75 Squadron painted this Hornet to signal the end of operations with the type.

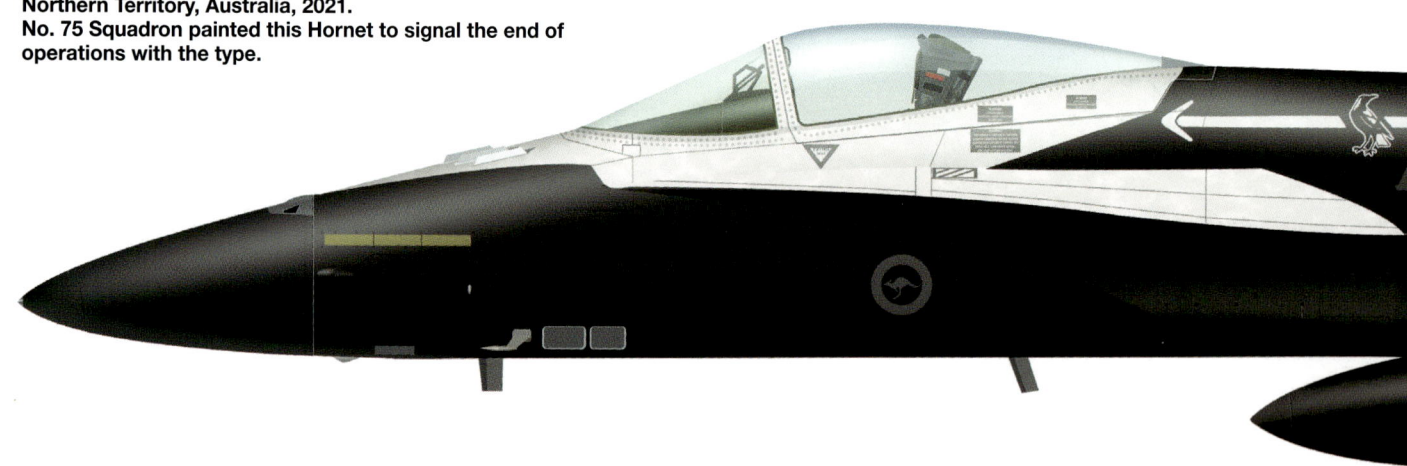

KUWAIT

Forty Hornets – 32 F/A-18Cs and eight F/A-18Ds – were ordered by Kuwait in August 1988. They had been scheduled for delivery in August 1991 but this was delayed when Iraq invaded. Following the Gulf War, deliveries commenced with the first three arriving in Kuwait on January 25, 1992. The last examples were delivered on August 21, 1993. Today, the Kuwait Air Force still operates 27 F/A-18Cs and seven F/A-18Ds.

FINLAND

Following a flyoff between the F/A-18C/D, F-16A, SAAB Gripen, Dassault Mirage 2000-5 and MiG-29, Finland announced on May 1992 that the Hornet had won. A total of 64 aircraft were ordered – 57 F/A-18Cs and seven F/A-18Ds – which would be assembled by the Valmet Aircraft Industry Company in Finland from McDonnell-prepared kits.

Finland intended its Hornets for air-to-air combat only, so they were designated F-18C and F-18D, dropping the '/A'. The first four F-18Ds were delivered from the US on November 7, 1995, with the first Valmet-assembled F-18C being delivered in June 1996. The last F-18C was delivered on August 8, 2000. Today, Finland continues to operate 55 F/A-18Cs and seven F/A-18Ds – the aircraft having since been restored to their fighter/ground-attack role.

SWITZERLAND

Having examined the Dassault Mirage 2000, Israeli Aircraft Industries Lavi, Northrop F-20 and SAAB Gripen, Switzerland narrowed down its choice for a new fighter in 1988 to the F-16 and F/A-18. Following a flyoff, the Hornet

came out on top and it was announced that 26 F/A-18Cs and eight F/A-18Ds would be acquired. Three years later the competition was reopened so that the Dassault Mirage 2000-5 and MiG-29 could be assessed. The F/A-18 defeated them both and the order for 34 F/A-18s was signed.

The first aircraft of each type would be supplied complete by McDonnell Douglas but the remaining aircraft would be assembled in Switzerland by Schweizer Flugzeuge und System AG at Emmen. The first F/A-18D for Switzerland flew for the first time at St Louis on January 20, 1996, with the first F/A-18C following on April 8, 1996.

The first Swiss-assembled F/A-18C flew for the first time on October 3, 1996, and the order was concluded with the delivery of last F/A-18C on December 2, 1999. Today Switzerland flies 25 F/A-18Cs and five F/A-18Ds.

MALAYSIA

The Malaysian Defence Minister announced the unusual decision, on July 1, 1993, to purchase both MiG-29s and F/A-18s. The letter of offer and acceptance on eight F/A-18Ds was signed on December 9, 1993, with 18 MiG-29s being ordered on June 7, 1994. The first F/A-18D was delivered on March 19, 1997, with the last having arrived by September of that year.

The Royal Malaysian Air Force F/A-18Ds went to war in 2013 when armed militants from the Philippines invaded Sabah on the island of Borneo – providing close air support for Malaysian ground forces. All eight remain in service today.

US JET FIGHTERS IN FOREIGN SERVICE — Boeing F/A-18E/F Super Hornet

BOEING F/A-18E/F SUPER HORNET

The jack-of-all-trades Super Hornet has so far been bought by two international operators – Australia and Kuwait.

2010–PRESENT

▼ BOEING F/A-18F SUPER HORNET
Boeing F/A-18F Super Hornet, A44-210, 1st Squadron, Royal Australian Air Force, Amberley air base, Queensland, Australia, 2016.
This Super Hornet carries tail art to celebrate the 100th anniversary of the squadron.

When the US Navy's McDonnell Douglas A-12 Avenger II programme was cancelled in 1991, it was decided that the next best option was to build on the reliable and combat-tested design of the F/A-18 Hornet – with bigger wings, a longer fuselage, new engines and new rectangular inlets.

McDonnell Douglas received a contract in June 1992 for seven prototypes of what would be designated F/A-18E and F/A-18F – the single and two-seater variants respectively. Its official name was Super Hornet, though it became known as the 'Rhino' in service to avoid confusion with its predecessor.

It was intended to replace the F-14 on fleet defence, the F/A-18C on air superiority, strike and reconnaissance, the S-3B as a tanker and the EA-6B on electronic warfare.

The first prototype or 'engineering and manufacturing development' aircraft made its flight debut on November 29, 1995, with testing commencing the following year and the type's first carrier landing taking place in 1997. Initial production began in March 1997 and full production began six months later. Following a series of corporate takeovers and mergers, the McDonnell Douglas F/A-18E/F became the Boeing F/A-18E/F. In 2003 an F/A-18F was heavily modified as a prototype for the electronic warfare version of the aircraft – the EA-18G Growler – and the first production examples were delivered in 2008.

The Australian government signed an agreement in 2007 to buy 24 F/A-18Fs as replacements for its F-111 fleet. The first five were delivered in March 2010 and the first Australian F/A-18F squadron was declared operational that December. A dozen EA-18Gs were then ordered in 2013 and at the time of writing 11 had been delivered. Eight RAAF F/A-18Fs took part in operations against Islamic State militants in 2014.

Kuwait signed a deal to buy 22 single-seat F/A-18Es and six two-seater F/A-18Fs in 2016. The 28 aircraft were reportedly built mostly to the US Navy's latest Block 3 standard but with some legacy Block 2 elements retained. All had been completed by September 2021 but were delivered to the US Navy, rather than going directly to the customer. After more than a year of training in the US, full delivery to Kuwait was expected to have been completed by the end of 2022.

▼ BOEING F/A-18F SUPER HORNET

Boeing F/A-18E Super Hornet, 803, Kuwait Air Force, Boeing Saint Louis facilities, Missouri, USA, 2021.
The first single-seat Super Hornet for Kuwait is seen here as it appeared during its first flight.

US JET FIGHTERS IN FOREIGN SERVICE — Lockheed Martin F-35A/B Lightning II

LOCKHEED MARTIN F-35A/B LIGHTNING II

Just when it seemed as though the days of huge US success in jet fighter export were over – along came the F-35. Despite questions being raised about its reliability and its technology, the stealth strike fighter has convinced many air forces around the world that its unique qualities are what they need.

2017–PRESENT

▼ LOCKHEED MARTIN F-35A LIGHTNING II

Lockheed Martin F-35A Lightning II, L-001, *Flyvevåbnet* (Royal Danish Air Force), Luke Air Base, Arizona, USA, 2021. The first RDAF F-35 was delivered in April 2021 and is based in Arizona for training; it carries full-colour markings.

The Joint Strike Fighter (JSF) programme emerged in 1993 from an amalgamation of earlier projects with the goal of creating an affordable strike fighter that would also be second only to the F-22 in the air supremacy role. Designs were submitted by McDonnell Douglas, Northrop Grumman, Lockheed Martin and Boeing with the latter two being chosen for further development.

Each company received a $750 million contract to develop prototypes on November 16, 1996, with Lockheed Martin's entry receiving the designation X-35 while the Boeing type became the X-32.

Two X-35 prototypes were developed – the X-35A, later converted into the X-35B, and the X-35C, which had larger wings. The X-35A completed its first flight on October 24, 2000, and the process of converting it into the X-35B commenced on November 22, 2000. The X-35C made its first flight on December 16, 2000. During final qualifying Joint Strike Fighter flight trials, the X-35B STOVL (short take-off, vertical landing) aircraft was able to take off in less than 500ft, go supersonic, then land vertically – which Boeing's equivalent design was unable to match.

Lockheed Martin was declared the winner and awarded a contract for system development and demonstration on October 26, 2001. The JSF programme was by now being jointly funded by the US, UK, Italy, Holland, Canada, Turkey, Australia, Norway and Denmark.

During further development, Lockheed Martin slightly enlarged the X-35, stretching the forward fuselage by 5in to make additional space available for the avionics. The tailplanes were correspondingly moved 2in further back to retain balance. The upper fuselage was raised by an inch along the centre line too. Parts manufacture for the first prototype began on November 10, 2003.

The X-35 had lacked a weapons bay and adding one resulted in design

▼ LOCKHEED MARTIN F-35A LIGHTNING II

Lockheed Martin F-35A Lightning II, A35-022, No. 3 Squadron, Royal Australian Air Force, RAAF Base Williamtown, New South Wales, Australia, 2021.
This Australian F-35 is seen with externally-carried GBUs during exercise Arnhem Thunder 21.

US JET FIGHTERS IN FOREIGN SERVICE — Lockheed Martin F-35A/B Lightning II

▼ LOCKHEED MARTIN F-35A LIGHTNING II

Lockheed Martin F-35A Lightning II, 32-05/MM7336, *XIII Gruppo, 32° Stormo, Aeronautica Militare* (13th Group, 32nd Wing, Italian Air Force), Amendola Air Base, Italy, 2017.
This F-35A received special tail markings to celebrate the 100th anniversary of several squadrons.

changes which increased the aircraft's weight by 2200lb. Lockheed Martin addressed this by increasing engine power, thinning the airframe members, reducing the size of the weapons bay itself and the size of the aircraft's fins. The electrical system also underwent changes, as did the section of the aircraft immediately behind the cockpit. All this succeeded in reducing weight by 2700lb but at a cost of $6.2 billion and 18 months of additional development time.

The F-35 is powered by a single 50,000lb thrust Pratt & Whitney F135 engine which, while it lacks a supercruise function, does enable the aircraft to fly at Mach 1.2 for 150 miles without using its afterburner. With afterburner, the F-35 has a top speed of Mach 1.6. The short take-off and vertical landing variant, the F-35B, has the Rolls-Royce LiftSystem. This features a thrust vectoring nozzle, allowing the main engine exhaust to be deflected downward from the tail end of the aircraft.

Base armament is a GAU-22/A 25mm cannon mounted internally and carrying 182 rounds in the F-35A or externally as a pod with 220 rounds for the F-35B and C. The pod itself has stealth features. All three F-35 variants have four underwing pylons capable of carrying AIM-120 AMRAAMs, AGM-158 cruise missiles and guided bombs. They also have two near-wingtip pylons designed for the AIM-9X Sidewinder and AIM-132 ASRAAM.

▼ LOCKHEED-MARTIN F-35I *ADIR*

Lockheed Martin F-35I *Adir*, 924, 5601 Squadron, Flight Testing Center, Israeli Air Force, Tel-Nof air base, Israel, 2020.
The first F-35I for the air force flight test centre was received in 2020; this unit is responsible for several programmes related to aircraft and weapons testing.

The F-35's two internal weapons bays can carry up to four weapons – two of them air-to-surface missiles or bombs and the other two air-to-air missiles such as the AIM-120 or AIM-132. Using both internal and external stations an air-to-air missile load of eight AIM-120s and two AIM-9s is possible.

Inside its cockpit, the F-35 has a 20x8in touchscreen, a speech-recognition system, a helmet-mounted display, a right-hand side stick controller, a Martin-Baker ejection seat and an oxygen generation system derived from that of the F-22. Due to the helmet display, the aircraft does not have a HUD. Its radar is the AN/APG-81 developed by Northrop Grumman Electronic Systems with the addition of the nose-mounted Electro-Optical Targeting System. The F-35's electronic warfare suite is the AN/ASQ-239 (Barracuda) with sensor fusion of radio frequency and infrared tracking, advanced radar warning receiver including geolocation of targeting of threats, and multispectral image countermeasures.

The aircraft has 10 radio frequency antennas embedded in its wings and tail. Six passive infrared sensors are distributed across the F-35 as part of Northrop Grumman's AN/AAQ-37 distributed aperture system. This provides missile warning, reports missile launch locations, detects and tracks approaching aircraft and replaces traditional night vision devices.

The first F-35, AA-1, was rolled out on February 20, 2006, and the type was formally given the name Lightning II on July 7, 2006 – the 'II' making it a spiritual successor to the Lockheed P-38 Lightning of the USAAF during the Second World War. AA-1 made its flight debut on December 15, 2006, and the first F-35B flew for the first time on June 11, 2008. A total of six F-35s had been built by January 5, 2009.

The first production F-35A, known as AF-6, flew for the first time on February 25, 2011.

BRITAIN

The UK was the earliest JSF partner – agreeing to pay $200 million to buy in to the project in November 1995. The following year, the UK's Ministry of Defence launched the Future Carrier Borne Aircraft project to find a replacement for the Sea Harrier. In 1997, UK defence contractor BAE Systems joined Lockheed Martin's team, alongside Northrop Grumman. Future Carrier Borne Aircraft was renamed Future Joint Combat Aircraft in 1998 and the Joint Strike Fighter was chosen for this role in 2001. As a Tier One partner,

US JET FIGHTERS IN FOREIGN SERVICE
Lockheed Martin F-35A/B Lightning II

▼ LOCKHEED MARTIN F-35A LIGHTNING II
Lockheed Martin F-35A Lightning II, 5209, *332 skvadron, Luftforsvaret* (332 Squadron, Royal Norwegian Air Force), Ørland Air Base, Norway, 2019.
Initial operational capability for the Norwegian F-35 was declared in 2019; this aircraft is shown with the weapons bay open.

the UK had an input on the aircraft's design and the selection of the prime contractor for the project.

It was announced in 2002 that both the RAF and the Royal Navy would operate the F-35B variant, with a tentative agreement to order 150 aircraft. Squadron Leader Steve Long became the first British pilot to fly an F-35 on January 26, 2010, and the UK ordered its first batch of 48 aircraft in July 2012. That same month, on the 19th, the first of these was delivered for trials work.

In 2015 the overall UK order was revised to 138 F-35B Lightnings – the 'II' being dropped in the UK, since it is actually the RAF's third Lightning after the Lockheed P-38 and English Electric Lightning. The first British F-35 unit, 617 Squadron 'Dambusters' became operational with the aircraft in June 2018. British F-35s will operate from the two Queen Elizabeth-class carriers – HMS *Queen Elizabeth* and HMS *Prince of Wales*. At present the UK has around 48 F-35Bs, with another 26 due for delivery, for a total of 74, with 138 now an 'upper limit' on potential future purchases.

ITALY
Joining the JSF programme in June 2002 as a Tier Two partner, Italy announced plans to buy 60 F-35As and 15 F-35Bs for the Italian Air Force, plus another 15 F-35Bs for the Italian Navy, making a total of 90 aircraft. As part of Italy's agreement with Lockheed Martin, F-35As and Bs are assembled at a facility in Cameri – the only F-35 production capability outside

▼ LOCKHEED MARTIN F-35A LIGHTNING II

Lockheed Martin F-35A Lightning II, F-001, 323 squadron, *Koninklijke Luchtmacht* (Royal Netherlands Air Force), Volkel air base, The Netherlands, 2019.
This special tail art, celebrating 70 years of the squadron, is seen during the RNLAF open day in 2019 *(Luchtmachtdagen 2019)*.

the US. The first Italian-produced F-35 made its flight debut at Cameri in 2015 and by September 2022, the Italian Air Force had received 17 F-35As and two F-35Bs, including three based in the US for training. The Navy had received three F-35Bs for a total of 22 F-35s.

NETHERLANDS
The other Tier Two partner, the Netherlands, intends to buy at least 46 F-35As. The Royal Netherlands Air Force received its first F-35A for testing in 2012 – stationed in the US – while the first Dutch F-35 squadron was stood up in 2018. Delivery of the first 24 F-35As was completed in January 2022.

AUSTRALIA
Australia signed up to the JSF programme in 2002 as a Tier Three partner and received its first F-35A in 2014. It currently intends to purchase 72 examples, with 54 having already been delivered by October 2022 leaving 18 still on order.

NORWAY
Another Tier Three partner in the JSF programme, Norway received its first F-35A in 2015 and had received a total of 37 by October 2022 with another 15 still on order – for an overall planned total of 52 examples.

ISRAEL
Having signed a letter of agreement in 2003, Israel joined the F-35 programme. A formal purchase of 20 aircraft was agreed in 2010, with the Israeli version to

US JET FIGHTERS IN FOREIGN SERVICE — Lockheed Martin F-35A/B Lightning II

be named the F-35I Adir (Adir meaning 'Awesome' or 'Mighty One'). There was also an option to purchase an additional 75. An order for 24 more was confirmed in 2014, then another 14 in 2015. The Israeli Air Force received its first F-35I in December 2016 and the first nine F-35Is were reported as operational by December 2017 – making Israel the first country outside the United States to have an operational F-35 squadron. During clashes with Iran in 2018, the IAF reported that it had become the first air force in the world to use the F-35 in combat, having used it to attack Iranian ground targets. These clashes continued into 2019 and on March 15, 2021 F-35Is shot down two Iranian drones carrying weapons to the Gaza Strip. This was the first operational shoot down and interception by an F-35.

As of October 2022, the IAF was operating 33 F-35Is with 17 more on order and plans to eventually have a fleet of 75.

DENMARK
A relative latecomer, Denmark decided to purchase the F-35 in 2016 as a Tier Three partner, with a total of 27 F-35As to be ordered. The first example was delivered in April 2021 – though it will remain in the US for training. The first operational Danish F-35s are due for delivery in 2023, with deliveries expected to continue until 2026.

CANADA
As the second international partner to join the F-35 programme after the UK, in 1997, Canada helped to fund the aircraft's development. In July 2010, it was announced that Canada would buy 65 aircraft – but there was growing opposition to the purchase. The F-35 then became a key issue during the Canadian federal election of 2011 and over the next two years a fierce debate raged. Boeing then stepped in and attempted to persuade the Canadians to buy the F/A-18 Super Hornet instead, with Dassault also trying to sell them its Rafale. The Liberal Party of Canada won the federal electron of 2015 with a promise not to buy the F-35. Three years later, Canada began a new competition to assess the Eurofighter Typhoon, F/A-18E/F, F-35 and Saab Gripen. Eventually, in March 2022, the Canadian government announced that the F-35A had won the contest and plans were set to buy 88 F-35A aircraft. If this now goes ahead, deliveries could commence in 2025.

▼ LOCKHEED MARTIN F-35A LIGHTNING II
Lockheed Martin F-35A Lightning II, 18-003, 151st Fighter Squadron, 17th Fighter Wing, Republic of Korea Air Force, Cheongju Air Base, South Korea, 2019.
South Korea plans to add the F-35B STOVL variant in the future, complementing its F-35As.

▼ LOCKHEED MARTIN F-35B LIGHTNING II
Lockheed Martin F-35B Lightning II, ZM150/016, 617 Squadron, Royal Air Force, HMS Queen Elizabeth, 2019.
RAF F-35Bs operated from the Royal Navy's aircraft carrier, alongside similar aircraft from the US Marine Corps.

JAPAN

Having joined the F-35 programme in 2011, Japan has the stated intention of purchasing 147 F-35s – the largest fleet outside the US. This would comprise 105 F-35As and 42 F-35Bs. Like Italy, Japan will build its own aircraft at a final assembly and checkout facility in Nagoya, established by Mitsubishi Heavy Industries, Mitsubishi Electric Company and IHI Corporation.

The first Japanese F-35 – built in the US – made its flight debut in 2016 and deliveries of operational aircraft to Japan commenced in 2018. At the time of writing, 23 F-35As had been delivered.

SOUTH KOREA

The Korean government signed a letter of offer and acceptance for 40 F-35As on September 30, 2014. The first RoKAF aircraft made its public debut in March 2018 and deliveries to Chongju Air Base commenced in 2019. In July 2022, it was announced that a further 20 would be purchased. All 40 aircraft from the first batch were said to have been delivered by March 2022, though since then concerns have been raised about their reliability.

BELGIUM

The government of Belgium signed up to purchase 34 F-35As in 2018. Deliveries are expected to commence in 2024.

SINGAPORE

After initially choosing to buy four F-35Bs with an option for another eight in 2019, it was revealed in September 2022 that Singapore was considering switching to another variant – either the A or C – before deliveries commence in 2026.

POLAND

It was announced on January 31, 2020, that Poland would buy the F-35. Thirty-two F-35As have been ordered and deliveries are expected to begin in 2024.

SWITZERLAND

The F-35A won a fighter competition held by the Swiss in 2021 and the government has said it intends to buy 36 examples, with the purchase order being signed in September 2022.

FINLAND

A letter of offer and acceptance for 64 F-35As was signed by the Finnish government in February 2022. Deliveries are expected to commence in 2025.

GREECE

Greece requested the purchase of 20 F-35As in July 2022. If the sale is approved, delivery would take place after 2028.

CZECH REPUBLIC

The Czech Republic entered negotiations with the US government to purchase 24 F-35As in July 2022. If the purchase goes through, deliveries are likely to begin during or after 2027.

GERMANY

The US state Department approved an $8bn deal for up to 35 F-35As for Germany in July 2022. Deliveries would begin in 2026 or 2027.

US Supersonic Bomber Projects

by Scott Lowther

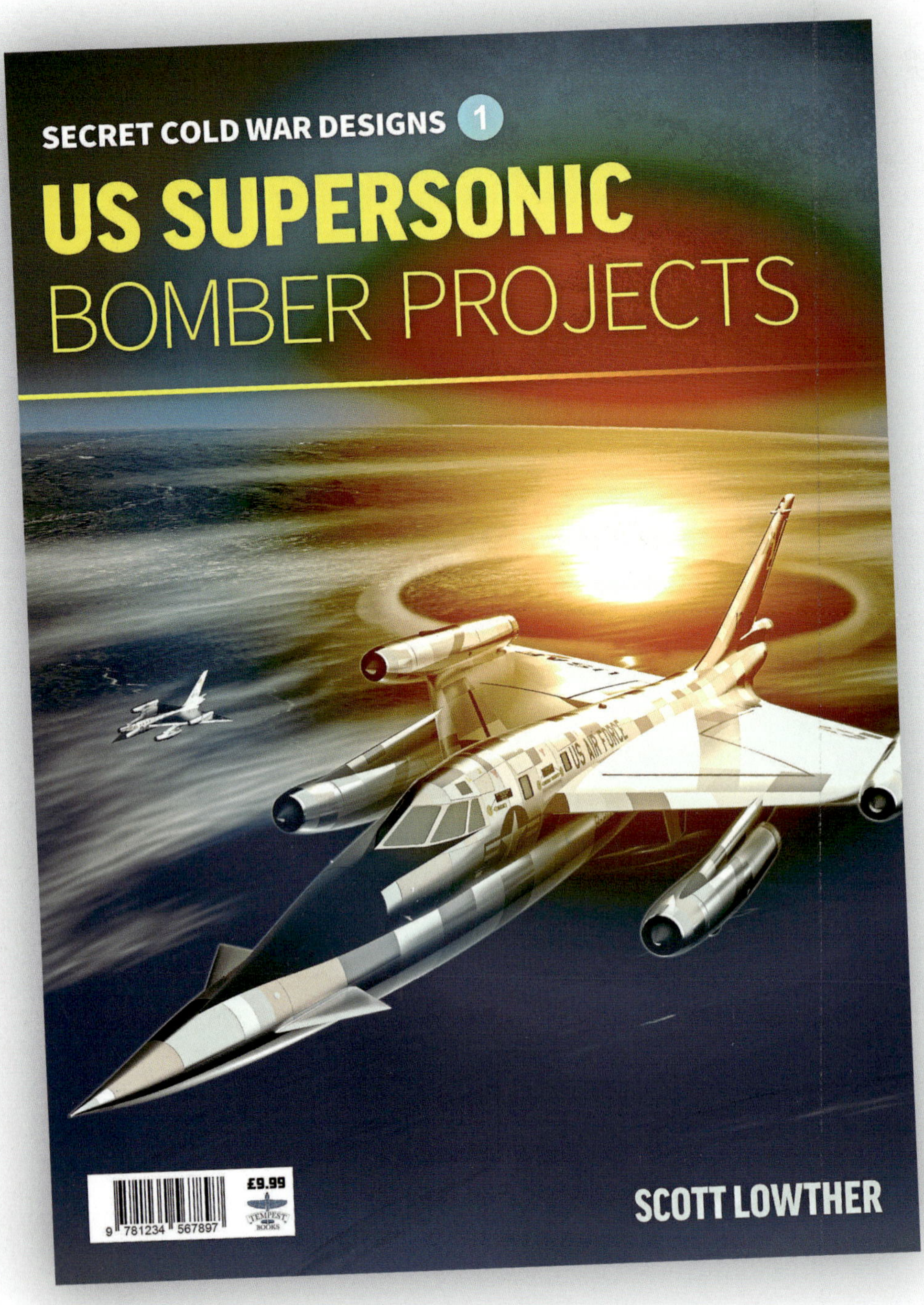

During the early stages of the cold war the USAF operated a huge fleet of strategic bombers - but these were all subsonic and development of high-speed surface-to-air missiles and jet interceptors made them increasingly vulnerable. America's Strategic Air Command urgently needed new aircraft capable of flying higher and faster than anything that the opposition could field.

The result was a vast number of supersonic bomber projects, ranging from viable proposals which became real aircraft, such as the B-58 and B-1, to incredibly ambitious concepts.

Using a wealth of illustrations, aerospace engineer Scott Lowther explores this unique period of aviation history in US Supersonic Bomber Projects.

ONLY £9.99
Quote: 5630

ORDER NOW:
www.classicmagazines.co.uk or call 01507 529529

Also by JP Vieira:

The US Air Force, Navy and Marine Corps all have a proud history of flying fast, well-equipped, and powerful jet fighters. Each of these bookazines features dozens of full colour profile artworks showing the very best of America's fighter force, created by renowned aviation illustrator JP Vieira.

USAF Fighters - Quote: 5578

Features more than 200 profiles including all 17 of the USAF's front line jet fighters.

US Navy Jet Fighters - Quote: 5590

Features more than 150 profiles including all 18 of the Navy's front line jet fighters.

ALL £8.99 EACH

Marine Corps Jet Fighters
- Quote: 15568

Having started out with McDonnell's basic FH-1 Phantom, the Marines soon received much more capable jets. This bookazine offers a fully illustrated history of these incredible aircraft with more than 150 highly detailed artworks.

ORDER NOW:
www.classicmagazines.co.uk or call 01507 529529

US JET FIGHTERS
IN FOREIGN SERVICE

Since the Second World War, American aircraft companies have built tens of thousands of aircraft not just for the USAF, US Navy and US Marine Corps, but also for America's friends and allies.

While some types were only sold or donated to foreign countries in relatively small numbers – such as Lockheed's pioneering F-80C, the McDonnell F2H-3 Banshee, Convair's F-102 Delta Dagger and, infamously, the Grumman F-14 Tomcat – others enjoyed massive export success.

Types such as the F-86 Sabre, F-4 Phantom and F-5 Freedom Fighter/Tiger II filled the hangars and airfields of air forces around the world. America's policy of aggressively seeking sales abroad has sometimes resulted in controversy. The Lockheed F-104 Starfighter, bought and even built in unprecedented numbers by European nations, became the focus of a bribery scandal during the 1970s.

Today American fifth generation types, such as the McDonnell Douglas (now Boeing) F-15 and F/A-18, and the General Dynamics (now Lockheed) F-16 continue to sell and the Lockheed F-35 Lightning II stealth strike fighter is already on track to join their ranks as a global success.

US Jet Fighters in Foreign Service offers a fully illustrated export history of these and many more incredible aircraft with more than 150 highly detailed artworks by renowned aviation illustrator JP Vieira included.